Organelles

To my parents

Our days are precious but we gladly see them going,
If in their place we find a thing more precious growing:
A rare, exotic plant, our gardener's heart delighting;
A Child whom we are teaching, a booklet we are writing.

Hermann Hesse
(1877–1962)

MOLECULAR CELL BIOLOGY
Series Editor: Dr C. J. Skidmore

Organelles

Mark Carroll, B.Sc., Ph.D.
Department of Biochemistry
The London Hospital Medical College
London E1 2AD, U.K.

The Guilford Press
NEW YORK   LONDON

Library of Congress Cataloging-in-Publication Data

Carroll, Mark.
    Organelles.

    (Molecular cell biology)
    Includes bibliographies and index.
    1. Cell organelles.   I. Title.   II. Series.
QH581.2.C38   1989   574.87′3   88-24625
ISBN 0-89862-403-7
ISBN 0-89862-526-2 (pbk.)

Printed in Hong Kong

CONTENTS

SERIES EDITOR'S PREFACE

The aim of this Series is to provide authoritative texts of a manageable size suitable for advanced undergraduate and postgraduate courses. Each volume will interpret a defined area of biology, as might be dealt with as a course unit, in the light of molecular research.

The growth of molecular biology in the ten years since the advent of recombinant DNA techniques has left few areas of biology unaffected. The information explosion that this has caused has made it difficult for large texts to keep up with the latest advances while retaining a proper treatment of the basics.

These books are thus a timely contribution to the resources of the student of biology. The molecular details are presented, clearly and concisely, in the context of the biological system. For, while there is no biology without molecules, there is more to biology than molecular biology. We do not intend to reduce biological phenomena to no more than molecular phenomena, but to point towards the synthesis of the biological and the molecular which marks the way forward in the life sciences.

C. J. Skidmore

ACKNOWLEDGEMENTS

In writing a book for the first time, I was helped along the way by numerous friends and colleagues. Dr Sue Danks provided the initial stimulus. Special thanks go to those who contributed micrographs, some of them previously unpublished: Dr Ted Katchburian, Prof. Rosemary Leech, Prof. K. Tanaka, Prof. Ralph Steinman, Prof. Eeva Therman, and Dr U. Abei. Others assisted with advice and constructive criticism: Dr Peter McCrorie, Prof. Geoff Duckett, Dr Peter Heathcote, and Dr Chris Skidmore.

The preparation of the figures was professionally performed by colleagues at the London Hospital Medical College: the medical illustrators (Sue McDonnell, Debbie Fenton and Ling Kemp), and the medical photographers (Ivor Northey, Melanie Tucker and Claire Lee). Barbara Hayes typed the initial manuscript with great diligence, Janet Guidal helped me greatly with the index and Mary Waltham and her colleagues at Macmillan Education patiently encouraged a novice author.

My wife, Sarah, and children, Samantha and Christopher, understandingly endured a rather unsocial husband/father as the writing proceeded; I can only hope that the end justifies the means.

Indeed, the final result is a book for which I take full responsibility. It will inevitably contain errors, whether of omission or commission. Like any good student, I should be pleased to hear from those who are more knowledgeable than I am.

London   Mark Carroll
January 1988

INTRODUCTION

● *Who will find this book useful?*

This book is intended for university or college students pursuing a biology, biochemistry or related course, for which cell biology is likely to be a significant component. However, some knowledge of elementary biochemistry is assumed: an understanding of the structure and function of the major biomolecules (proteins, lipids, carbohydrates and nucleic acids), of enzyme action (and coenzymes), of the major metabolic pathways, and of simple bioenergetics and genetics. Other potential users of the book will be those teachers, at university and college level, who need to be familiar with recent advances in cell biology. Indeed, some schoolteachers may also benefit from it, as the molecular aspects of cell biology are increasingly emphasised in the science curriculum.

● *How has cell biology evolved?*

Cell biology has come a long way since the first description of cells by Hooke and Leeuwenhoek in the 17th century. Their home-made light microscopes have been superseded by modern electron microscopes which can look into the heart of a single cell. Complementary to the imaging techniques of light microscopy and electron microscopy are the modern methods of biochemical analysis and molecular biology. We can dissect the components of the living cell, separate one from the other, study their form and function and how they arise. A reductionist approach such as this must necessarily be followed by a synthesis: an attempt to understand how the components interact, and how their activity is regulated by the cell. This philosophy is the essence of cell biology.

● *What are organelles?*

But what *is* cell biology? One of the great advances in evolution came with the development of **eukaryotes,** as opposed to the apparently more primitive **prokaryotes.** The former became relatively larger,

Eukaryotes: Unicellular and multicellular organisms whose cells have a true nucleus.
Prokaryotes: Unicellular organisms, mainly bacteria, whose cells lack a true nucleus.

and developed a subcellular division of labour that was associated with the generation of membrane-bound *organelles*. In fact, if one looks at the ultrastructure of a eukaryotic cell, one sees a complex panoply of membranous structures, each presumably with its own specialised functions and suspended in an apparently amorphous fluid, the *cytosol*. Although the numbers of these organelles and their morphology may vary somewhat from one organism to another, and from one tissue to another, the pattern remains remarkably constant. This is not surprising, for all eukaryotic cells have to carry out certain common functions: synthesis of ATP by oxidative phosphorylation; degradation of biopolymers such as proteins and polysaccharides; the covalent attachment of sugar residues to newly synthesised proteins; the disposal of potentially lethal oxidising agents; and of course the replication of the nuclear material (DNA) that accompanies cell division. In addition, the plant kingdom has an extra set of organelles, the plastids, which are associated mainly with their ability to harness the energy of sunlight in the process of photosynthesis. Cell biologists study all these organelles, as well as the interactions between those cells that make up the complex structures of multicellular organisms.

● *What are the disadvantages of organelles?*

Although the functional specialisation of discrete organelles within eukaryotic cells has obvious advantages, in terms of a close association of related components that can enhance productivity, there are also potential drawbacks. In a sense, an organelle represents an artificial barrier to metabolism. For example, the complete oxidation of glucose involves the operation not only of glycolysis but also of the tricarboxylic acid (Krebs) cycle. These two pathways are located in two distinct compartments of the cell. The fact that each organelle is usually surrounded by a limiting membrane, which is inherently impermeable to polar solutes, means that there is rarely free exchange of materials between different organelles, or between an organelle and the cytosol. Thus, cells have evolved elaborate mechanisms to facilitate the passage of certain solutes across their intracellular membranes. However, the evolutionary primacy of eukaryotic organisms implies that the advantages to be gained from this specialisation at the subcellular level more than outweigh the disadvantages.

● *What questions are addressed by cell biology?*

With the development of new biochemical and microscopic techniques, we have gained a reasonably clear picture of the structure and function of the cell's organelles. The excitement that cell biologists now feel has come about as a result of their ability to probe new aspects of their subject by applying these recently developed techniques to questions such as: how do new organelles arise? what are the structural and functional relationships between them? how does the DNA regulate the coordinated activity of the cell's organelles? what are the causes and effects of genetic lesions that disrupt the normal functioning of the cell?

● *What recent technical advances have contributed to the study of cell biology?*

Three areas of technical improvement have enhanced our understanding in recent years, and will continue to do so for the foreseeable future. Firstly, the development of *recombinant DNA techniques* has allowed us to clone individual genes, to determine their fine structure and, ultimately, the mechanisms by which they are controlled. From the base sequence of a structural gene one can derive the amino acid sequence of the encoded protein, and from there make predictions concerning potentially significant functional regions of the molecule. Secondly, we now have the ability to generate *monoclonal antibodies* that are exquisitely specific for a single site on the corresponding antigen. Such antibodies can be exploited to detect and to quantitate individual proteins and other cellular macromolecules. Thirdly, there have been considerable advances in the techniques of *electron microscopy* (EM), for we are no longer limited to the two-dimensional images provided by conventional transmission EM. Furthermore, individual components can now be identified at the subcellular level by immuno-gold methods, which represent the combination of monoclonal antibodies with EM. Examples of all three classes of technique will be found throughout the book.

● *What is the organisation of this book?*

The method of presentation of the material in this book has been chosen so as to focus attention on one particular organelle at a time. Thus, there are individual chapters on the nucleus, the endoplasmic reticulum, the Golgi complex, lysosomes, peroxisomes, mitochondria and chloroplasts. The order of chapters is such as to illustrate the flow of genetic information from the nucleus, *via* protein biosynthesis, to all organelles within the cell. For reasons of space, there is little consideration of the plasma membrane, despite its obvious importance in regulating the interaction of the cell with its environment. There is also only a brief treatment of some organelles and macromolecular structures, either because they are restricted to certain specialised cells, or because they were felt to fall outside the terms of reference of this book, which concentrates on membrane-bound organelles. One consequence is that there is only a superficial look at the ultrastructure of the cytosol. One other criticism that might be levelled at the presentation adopted in this book is, that it leads to an artificial compartmentalisation. In order to minimise this potential drawback, each chapter contains extensive references to related aspects elsewhere in the book. There is also a final chapter that serves to emphasise the point, with the help of selected examples, that there is considerable interplay between organelles.

To enhance the continuity of the text and students' understanding, the definitions of some possibly unfamiliar terms have been placed in the margin at the point when they are first used. Abbreviations which are used throughout the text are listed in appendix A at the end of the book.

● *What experimental methods are referred to in this book?*

The accent throughout the book is on the ways in which information derived from various experimental approaches has enhanced our understanding of subcellular structure and function. To this end, it was considered worthwhile to include a chapter describing the principles of the techniques that cell biologists use in order to probe the cell. However, chapter 2 is primarily for reference purposes, to be consulted when an unfamiliar experimental method is encountered in the text. In is *not* intended that students should read and inwardly digest this chapter! In particular, there are condensed accounts of gene cloning, monoclonal antibodies, and various methods of electron microscopy. The latter can provide beautiful images of the subcellular world. These images are not only aesthetically pleasing, however, for they can also improve our ability to formulate models that describe what is happening at the molecular and subcellular level. For this reason, it was considered important to include a considerable number of electron micrographs in order to illustrate and complement the text.

● *How can students test their understanding?*

The author feels strongly that students do not necessarily learn much from passive reading; they are more likely to consolidate their understanding if they participate actively in trying to manipulate the material which they have encountered. To this end, each chapter concludes with a succinct summary of its content, and a series of study questions relating to the material presented in that chapter. In some cases, the questions test simple recall or understanding, in others students are encouraged to extrapolate to other related systems. Brief answers to the study questions are presented in appendix B.

● *What additional reading is desirable?*

It is obvious that in a book such as this, the author has had to restrict the amount of information presented. As already mentioned, an understanding of the structure of biomolecules and of the common metabolic pathways has been assumed; these topics are adequately covered in standard biochemistry textbooks. There are also more specialised texts that cover in detail the molecular aspects of subcellular structure; these too are referred to in the general reading list at the end of this Introduction. In individual chapters there are suggestions for further reading. Invariably these references are not to the primary literature, but rather to more general reviews published in journals to which most students should have access.

● *What are the aims of this book?*

In conclusion, the prime aim of this book is to present an account of our current understanding of subcellular organelles at the molecular level: their structure, function, biogenesis, and interaction. The principles presented here should provide the student with an adequate basis to enable him or her to delve further into the subject, as well as to appreciate new developments as they arise. These are exciting times for cell biology — this book attempts to capture that excitement.

Further reading

(a)   General textbooks

There are many excellent general textbooks of biochemistry, including the following:

Stryer, L. (1988). *Biochemistry*, 3rd edn, W. H. Freeman and Co., San Francisco.
    (Superbly illustrated)
Lehninger, A. L. (1982). *Principles of Biochemistry*, Worth Publishers Inc., New York.
    (Written by one of the foremost researchers on mitochondria)

(b)   Specialised textbooks

Two accounts of cell biology are impressive for their clarity and comprehensiveness:

Alberts, B., Bray, D., Lewis, J., Raff, M., Roberts, K. and Watson, J. D. (1983). *Molecular Biology of the Cell*, Garland Publishing Inc., New York.
Darnell, J., Lodish, H. and Baltimore, D. (1986). *Molecular Cell Biology*, W. H. Freeman and Co., San Francisco.

An excellent source of electron micrographs of subcellular organelles is:

Fawcett, D. W. (1981). *The Cell*, 2nd edn, W. B. Saunders Co., Philadelphia.

1 ULTRASTRUCTURE OF THE CELL

1.1 The view through the light microscope

The term 'cell' was first coined by **Hooke** in 1665 to describe the compartments in cork that he could see through rather crude compound lenses. **Leeuwenhoek** improved the construction of the light microscope, such that in 1674 he was able to view protozoa, and subsequently bacteria. Developments in theory, design and construction allowed microscopists by the end of the nineteenth century to move towards the theoretical limit of their instruments: a **resolution** of 0.2 μm. With such instruments, they were able to describe the vast range of **morphology** of eukaryotic cells, the association of specialised cells into tissues and organs, and some of the subcellular architecture of the cell.

Resolution: The ability to discriminate, or visually separate, two adjacent points.
Morphology: Shape and size of an object.

1.1.1 Cells vary enormously in their size and shape

There is no 'typical cell'. Eukaryotic cells range in size from the egg cell of the ostrich (10^5 μm in diameter, with an approximate volume of 5×10^{14} μm^3) to a yeast cell (10 μm in diameter, with an approximate volume of 5×10^2 μm^3). They may be roughly spherical in shape, such as a mammalian ovum, or they may have an extremely elongated form, such as certain neuronal cells. They may have a smooth exterior, as in the mammalian erythrocyte, or they may have a highly convoluted surface, as in the mammalian leukocyte. They may function in isolation from other cells, as with the two blood cells just mentioned, or they may associate with other cells in order to generate functionally specialised tissues and organs. They may have surface structures essential for their function: the ruffled membranes of cells with cilia and pseudopodia, for example, or the motile flagella of spermatozoa and trypanosomes. They may be long-lived (like certain cells of the immune system, that may last the life-time of the individual), or be relatively short-lived (with a half-life of 120 days, like that of the erythrocyte). They may take on a huge variety of shapes: spherical, ovoid, biconcave disc, flattened, pointed, highly elongated, to mention just a few. Such is the great diversity of morphology among eukaryotic cells.

1.1.2   Light microscopy reveals the essential features of the cell

The concept of the cell as the organising principle behind the structure of tissues and organs was first proposed by **Schwann** in 1835. Since an average cell has a diameter of about 20 μm, it was readily accessible to the early microscopists. Some idea of what they saw is shown in figure 1.1, which is a **light micrograph** of an animal cell. Three structures are apparent: a limiting membrane (the *plasma membrane*) that defines the boundary of the cell: a centrally placed, relatively large *nucleus*; and the remaining contents of the cell, the *cytoplasm*, which generally appears as an amorphous, granular fluid. At this relatively low level of magnification, other subcellular organelles are not apparent in most cells. (One exception to these general observations is the mammalian erythrocyte, or red blood cell, which lacks a nucleus.)

Owing to its limited resolution, light microscopy was restricted to descriptions of the morphology of whole cells in various tissues, and in some cases the varied appearance of the nucleus, as occurs in the polymorphonuclear leukocyte, for example. Indeed, the nucleus was known to play an important role in cell division, a process which leads to gross changes in nuclear structure and the duplication of the organelle for transmission to each of the two daughter cells. However, the real significance of the contents of the nucleus only became apparent

Micrograph: A photograph taken through a microscope.

Figure 1.1
Light micrograph of an animal cell

This living human fibroblast cell has been observed under the light microscope with Nomarski optics. The prominent central nucleus (N) is surrounded by granular cytoplasm (CY), which in turn is delimited by the plasma membrane (PM). The movement of the cell has resulted in numerous projections (pseudopodia) and 'ruffles' in the plasma membrane. × 1000. [Courtesy of E. Katchburian and P. Purkis.]

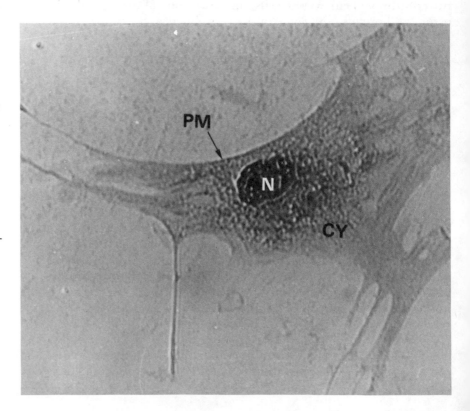

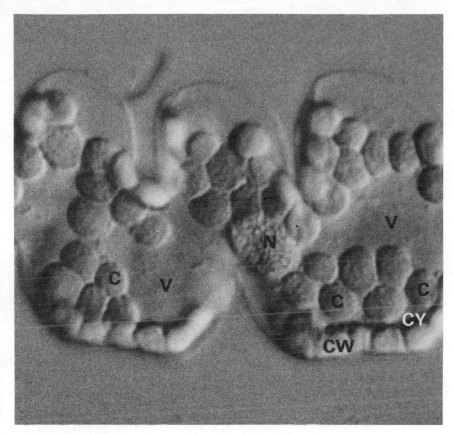

Figure 1.2
Light micrograph of a plant cell

In this living mesophyll cell from a wheat leaf (*Triticum aestivum*), the morphology is clearly visible when viewed by Nomarski optics. Note in particular the central nucleus (N), the thick cell wall (CW), the vacuoles (V), and the cytoplasm (CY) packed with chloroplasts (C), the organelles of photosynthesis. × 1000. [Courtesy of A. J. Jellings and R. M. Leech.]

much later. Furthermore, the function of the cytoplasm was also unclear in the nineteenth century.

The early microscopists also noted that plant cells had a somewhat different morphology (figure 1.2). Although they too possessed a plasma membrane, a nucleus and cytoplasm, they had two apparently unique structures: a dense *cell wall* around the cell, and large internal *vacuoles* surrounded by a *tonoplast membrane*. These features were readily explicable: higher plants are immobile organisms that require well-developed supporting structures. Their cells lay down an extensive extracellular network of cellulose and other polysaccharide material to form a cell wall that provides much of the necessary support. Plant cells stretch outwards against their walls owing to the turgidity endowed by the vacuole, which may comprise 90 per cent of the cell's volume and which contains a solution of sugars, amino acids, ions and proteins. With modern light microscopes, the photosynthetic organelles (*chloroplasts*) of plant cells are also visible (figure 1.2).

1.1.3   Certain stains highlight subcellular organelles

Under normal circumstances, the cell and its contents are translucent and hence not readily visible in the light microscope. Cells have to be fixed and stained before they become clearly identifiable (figure 2.1), unless phase-contrast microscopy is employed. Some of these stains were found to be selective for certain subcellular structures. In particular, the nucleus was clearly differentiated by some so-called vital dyes, and a densely staining intra-nuclear *nucleolus* could be discerned. Other dyes appeared to stain tiny oval or at times thread-like structures called *mitochondria*. A silver stain was found by **Golgi** to reveal the small membranous complex that bears his name. Thus it was that the early microscopists showed that the nuclear material and the cytoplasm might have some underlying structure. The fine detail of this structure would not be revealed, however, until the development of the *electron microscope*.

1.2   The view through the electron microscope

In the 1930s, **Ruska** developed methods for focusing beams of electrons, which have the wave structure of light (beams of photons) but a much shorter wavelength (section 2.1.2). The result was an electron microscope with a greatly enhanced resolving power, down to 2 nm or so. (Contrast the 200 nm resolution of the light microscope.) Such machines were able to uncover much more of the cell's internal complexity.

1.2.1   The electron microscope reveals the ultrastructure of the cell

Although the early electron microscopists did not have optimal techniques for fixing and staining cells, for cutting ultra-thin sections, for focusing the electron beam or for recording their results, nevertheless they were able to produce pictures of stunning complexity (figure 1.3). The nucleus was found to be surrounded by a double membrane, the *nuclear envelope*, pierced in places by *nuclear pores*. The mitochondria also had a double membrane, the inner one of which was extensively folded into *cristae*. The *Golgi complex* was revealed as a stacked array of flattened membranous sacs. The term *organelle* was coined to describe these membrane-bound subcellular structures.

In addition to these already known organelles, there were many others. An extensive network of membranes permeated the entire cytoplasm — the *endoplasmic reticulum*. In some cases these membranes were smooth, in others they were studded with *ribosomes*, later shown to be protein-synthesising structures. There were numerous *vesicles* (or *vacuoles*) with a single limiting membrane; some of these were clear (electron-lucid), others were referred to as *dense bodies* or, in certain cells, *secretory granules*. In photosynthetic plant cells furthermore (figure 1.4), there were characteristic organelles, the chloroplasts, with their double envelope and their arrays of internal membranes. Here was the site of photosynthesis, which during the hours of daylight

Figure 1.3
Electron micrograph of an animal cell (see opposite page)

In this thin section of a cell (hepatocyte) from rat liver, the major organelles are readily visible: N, nucleus; Nu, nucleolus; M, mitochondrion; GA, Golgi complex (apparatus); Ci, cisternae of the endoplasmic reticulum; L, lysosome; P, peroxisome; Gl, glycogen particles scattered throughout the cytosol. The plasma membranes (PM) of adjacent cells lie below the nucleus. (Not all the organelles are labelled.) Scale bar = 1 µm. [Taken from Cardell, R. R. (1971). *Amer. J. Anat.*, **131**, 21–54; reproduced by permission of the publisher, Alan R. Liss Inc.]

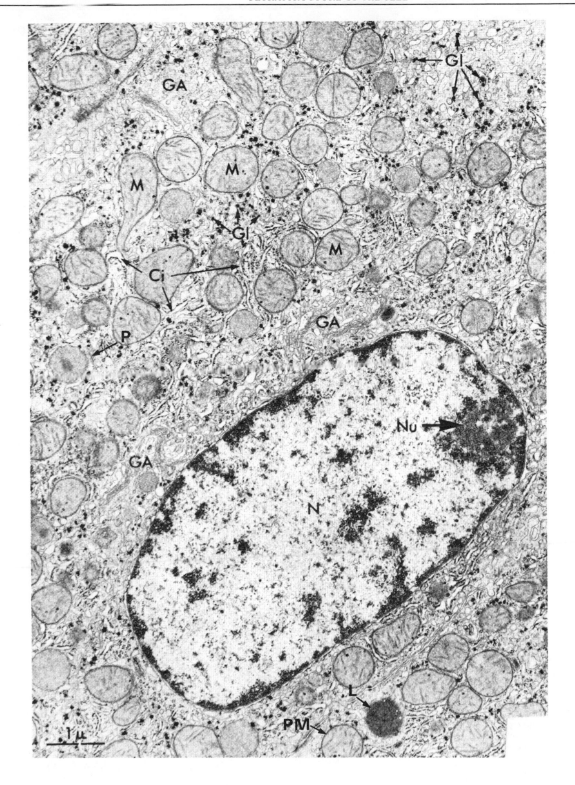

provides the plant cell with most of its ATP and reducing power needed for biosynthetic reactions.

All these organelles were suspended in an aqueous environment the so-called **cytosol**. (As we shall see in section 1.4.2, even the cytosol has an organised structure.) Furthermore, they were seen to be bounded by either a single or double membrane, with the exception of the ribosome. There were other non-membranous structures too, such as centrioles, lipid droplets, and granules of starch or glycogen. The *centrioles* existed as a pair of cylindrical structures close to the nucleus. In those cells which possess them, they are thought to orchestrate the apparatus (the mitotic spindle) that partitions the newly duplicated DNA between the daughter cells at cell division (mitosis). They are also thought to organise the *microtubules* which ramify throughout the cytosol. Limitations of space prevent us from considering these non-membranous organelles.

Cytoplasm/cytosol: The cell contains a nucleus and cytoplasm. If all the organelles were removed from the cytoplasm, the resulting aqueous fluid would be the cytosol.

1.2.2   Biochemical techniques complement electron microscopy

Electron microscopy can obviously tell us what the ultrastructure of the cell looks like, but it gives us little information on what the organelles actually *do*. It also presents us with a static picture of the cell's contents, whereas we know that the cell is a dynamic entity. What we need is the extra information that biochemistry can provide on the subcellular organisation of cellular function.

A major advance came about with the development of the *ultracentrifuge,* which provided biochemists with the ability to separate relatively large quantities of organelles from isotonic homogenates of tissues (section 2.2.1). In particular, **de Duve** in the 1950s refined the techniques of differential centrifugation and density-gradient centrifugation in order to characterise membrane-bound organelles associated with hydrolytic activity (*lysosomes*) or oxidative activity (*peroxisomes*). The discovery of these new populations of organelles was confirmed by their identification with the electron microscope. In fact, they could be assigned to subcellular structures described by electron microscopists years before, but whose function had not at that time been determined: the dense bodies (lysosomes) and the microbodies (peroxisomes), visible in figure 1.3.

Exocytosis: Fusion of a vesicle with the plasma membrane so as to release the contents of the vesicle into the extracellular space.

A further example of the way in which electron microscopy was complemented by biochemical approaches came about through the development of *autoradiography* (section 2.1.2). **Palade** studied the secretion of digestive enzymes by pancreatic cells, by labelling their newly synthesised protein with radioactive amino acids and tracing the path of the labelled proteins through the cell. The protein was first assembled on ribosomes associated with the rough endoplasmic reticulum, then migrated to the Golgi complex, where it was packaged into membrane-bound vesicles that developed into secretory granules, to be released from the cell by **exocytosis.** We shall see later in this book numerous examples of how cell biologists have harnessed imaging

Figure 1.4
Electron micrograph of a plant cell (see opposite page)

This thin section of a leaf cell from *Atriplex* species illustrates the characteristic ultrastructural features of photosynthetic cells: plastids (P; here, immature chloroplasts), vacuoles (V), and the cell wall (CW). The other organelles are also present in animal cells. × 10 000. [Original micrograph by W. W. Thomson; courtesy of R. M. Leech.]

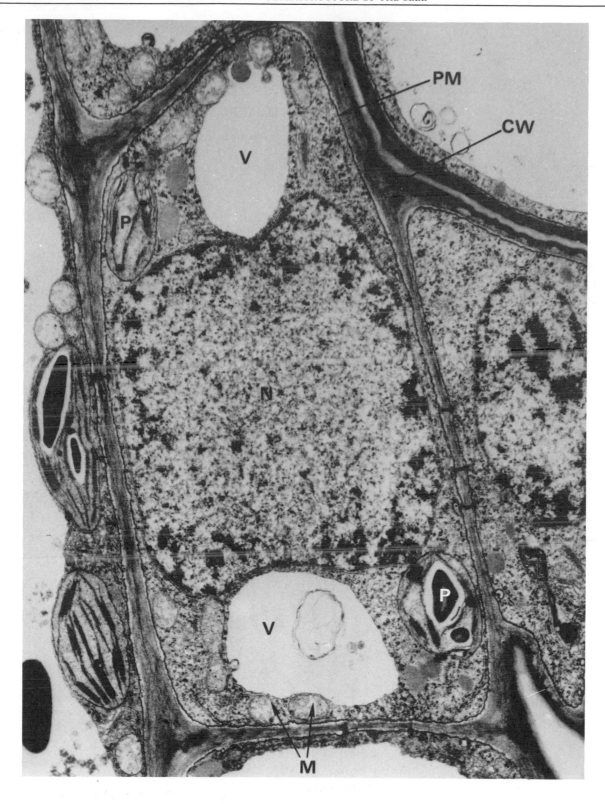

techniques, such as electron microscopy, to analytical (biochemical) techniques in order to study the structure and function of organelles.

1.3   Membranes divide the cell into compartments

The organelles which interest us here have a limiting membrane, which has an important structural and functional role to play. At this point, we need to consider briefly the organisation of biological membranes and some of their properties.

1.3.1   The fluid-mosaic model describes the structure of biological membranes

The currently accepted model of membrane structure was first proposed by **Singer** and **Nicolson** in 1972. Membranes are viewed as two-dimensional sheets of **amphipathic lipids** into which the inserted oriented globular proteins — the so-called *fluid-mosaic* structure (figure 1.5).

The essential features of this model are:

(a)   amphipathic lipids form a bilayer stabilised largely by hydrophobic interactions;

(b)   globular proteins associated with the lipid bilayer are either extrinsic (peripheral) or intrinsic (integral);

Amphipathic lipid: One with both a hydrophobic (non-polar) portion and a hydrophilic (polar) portion. For example, phosphatidylcholine (lecithin):

$$CH_2 OCO(CH_2)_n CH_3$$
$$CHOCO(CH_2)_n CH_3$$
$$CH_2 OPOCH_2 CH_2 \overset{+}{N}(CH_3)_3$$

Here, the long fatty acyl chains $[CH_3(CH_2)_n -]$ on C-1 and C-2 hydrophobic, whereas the phosphorylcholine moiety on C-3 is hydrophilic, in aqueous solution at neutral pH.

Figure 1.5
Fluid-mosaic model of membrane structure

Amphipathic lipids (phospholipid, glycolipid, cholesterol) form a thermodynamically stable bilayer, with the polar 'head' groups facing the aqueous phase and the fatty acyl 'tails' projecting internally. Globular proteins are either inserted into the bilayer (integral, or intrinsic, membrane proteins), or are bound to the surface (peripheral, or extrinsic, membrane proteins). Carbohydrate is associated with the extracellular face (OUT) of the plasma membrane.

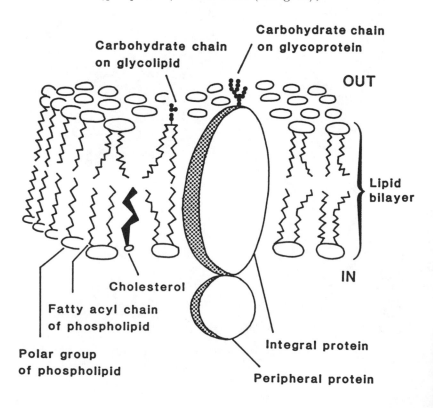

Carbohydrate chain on glycolipid

Carbohydrate chain on glycoprotein

OUT

Lipid bilayer

IN

Cholesterol

Fatty acyl chain of phospholipid

Polar group of phospholipid

Integral protein

Peripheral protein

(c) there is relatively free lateral mobility of components in the plane of the membrane, but highly restricted transverse mobility from one side of the bilayer to the other;

(d) the lipid bilayer is a fluid structure, with the fluidity being regulated by the number of double bonds in the fatty acids (increasing unsaturation increases fluidity) or the proportion of cholesterol (increasing cholesterol decreases fluidity at $37°C$);

(e) the lipid bilayer represents a permeability barrier to polar solutes and ions, which can only cross the membrane if transported by a specific membrane-bound protein;

(f) the components of biological membranes, and in particular the proteins, are asymmetrically disposed (oriented) within the membrane.

What are the implications of these properties for subcellular membranes?

1.3.2 Compartmentalisation within cells enhances function

We saw in the Introduction how eukaryotic cells are distinguished in part from prokaryotic cells by their vast array of internal membranes. These in turn create numerous compartments within the cell, such that each compartment represents an organelle separated from its neighbours by at least one membrane. Let us at this stage refresh our memories as to what these organelles are, and some of their characteristics (table 1.1). What advantages are gained by such a development?

Table 1.1
Some properties of the major compartments within the cell

Compartment	Approximate number per cell	Percentage of total cell volume	Percentage of total cell membrane
Cytosol	1	54	0
Nucleus	1	6.0	0.2
Endoplasmic reticulum:	1	—	—
rough	—	9.0	35
smooth	—	4.0	16
Golgi complex	1	1.5	7.0
Lysosomes	300	1.0	0.4
Peroxisomes	400	1.0	0.4
Mitochondria:	1700	22	—
inner membrane	—	—	32
outer membrane	—	—	7.0

The values quoted are for a 'typical' mammalian cell (hepatocyte). The **cisternae** of the rough and smooth endoplasmic reticulum are thought to form a single compartment, as do those of the Golgi complex. The total cell volume of a hepatocyte is approximately 5000 μm^3; the total cell membrane has an area of approximately 1.1×10^5 μm^2. [Adapted from Alberts, B. *et al.* (1983). *Molecular Biology of the Cell*, with permission of Garland Publishing Inc.]

Cisterna (plural, **cisternae**): Flattened membranous sac enclosing an aqueous space, the *lumen* or *matrix*.

Table 1.2
Organelles carry out specific metabolic functions

Organelle	Metabolic function
Nucleus	Synthesis of DNA and RNA
Golgi complex	Modification of glycoproteins
Lysosomes	Degradation of biopolymers
Peroxisomes	Reactions involving H_2O_2
Chloroplasts	Fixation of CO_2 by photosynthesis

This is only a representative sample of organellar functions; many more are discussed in specific chapters.

One consequence is that eukaryotic cells have a greatly expanded surface area for membrane-associated structures and their corresponding functions, which can then proceed at greatly enhanced rates. Another consequence is the formation of numerous membrane-bound vesicles, within which can be segregated those components that are functionally related. Thus, hydrolytic enzymes are sequestered within lysosomes (chapter 6), where they cooperate in the complete degradation of biological macromolecules. Again, the highly folded cristae of mitochondria (chapter 8) contain membrane-bound components that interact to form the electron transport chain involved in oxidative phosphorylation. There are numerous examples of how certain organelles are associated with specific metabolic functions (table 1.2).

There is one further consequence of the fact that all biological membranes are organised according to the same underlying principles: membranes are able to fuse with one another. In this way, the membrane components of one cell (or vesicle) are able to mingle with those of another, and likewise the soluble components. Clearly membrane fusion is not a random event, but it is essential for certain cellular functions, such as the lysosomal degradation of macromolecules or the exocytotic release of secreted digestive enzymes and polypeptide hormones.

Biogenesis: The mechanism by which new organelles, or other subcellular structures, arise.

The implication of such subcellular organisation is that the cell is able to incorporate specific lipids and proteins into a specific class of organelle. How does it achieve this goal? The **biogenesis** of organelles (section 1.5.1) and the assembly of their component parts is one of the major challenges facing cell biologists today.

1.3.3   Subcellular membranes represent permeability barriers

There is one drawback, however, to this compartmentalisation of metabolism and that concerns the restrictions which are placed on the free movement of metabolites by the presence of semipermeable intracellular membranes. By their very nature, biological membranes are not freely permeable to polar solutes (section 1.3.1). So for example, pyruvate generated by glycolysis in the cytosol must gain access to the interior of the mitochondrion before it can be completely oxidised. Hence, the inner mitochondrial membrane contains a transport protein that admits the pyruvate. The entry of the negatively charged pyruvate must be balanced by the simultaneous entry of a positively charged solute, or the exit of a negatively charged species; in fact, electrical neutrality is maintained by the transport protein swapping the pyruvate for a hydroxyl ion (OH^-). This example further illustrates another general principle: membrane proteins are invariably oriented so as to achieve the vectorial (unidirectional) transport of a specific solute.

In some cases, there is no appropriate transport protein. For example, NADH generated by glycolysis in the cytosol cannot cross the inner mitochondrial membrane in order to gain access to the electron transport chain. Instead, the reducing equivalents (hydrogen atoms, or electrons) carried by NADH have to be ferried across the membrane by a rather complex *shuttle system* (figure 1.6). There is a metabolic price to pay for the efficient operation of this system, however: one ATP is expended per pair of electrons transported.

There are certain terms associated with membrane transport that we will use throughout this text; they are presented in the margin. Do not confuse membrane transport (the passage of solutes across a membrane) with membrane flow (the intracellular movement of the membranes themselves). The processes which bring about membrane

Transport protein: Membrane-bound protein that facilitates the passage of (usually) one specific solute across a membrane (also known as *portery* or *permeasey*).
Uniport: Unidirectional passage of one solute across a membrane.
Antiport: Simultaneous transport of two solutes (usually with similar charges) in opposite directions across a membrane.
Symport: Simultaneous transport of two solutes in the same direction across a membrane.

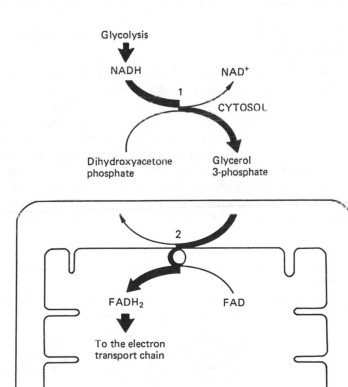

Figure 1.6
A shuttle system in metabolism

The reactions indicated are catalysed by a cytosolic NAD^+-linked dehydrogenase (1) or a mitochondrial FAD-linked dehydrogenase (2). The overall effect of the operation of this shuttle is to transfer reducing equivalents (hydrogen atoms, or electrons) from the cytosol into the mitochondrion for oxidative phosphorylation *via* the electron transport chain. The path of the electrons is traced by the bold arrows.

Table 1.3
Characteristics of membrane transport processes

Property	Passive diffusion	Facilitated diffusion	Active transport
Saturation at high substrate concentration	No	Yes	Yes
Protein mediator	No	Yes	Yes
Energy requirement	No	No	Yes
Against a concentration gradient	No	No	Yes

transport are defined as *passive diffusion, facilitated diffusion,* or *active transport.* Some of their properties are presented in table 1.3.

1.4 The cytosol is a structured aqueous compartment

The organelles that we have described are suspended in the cytosol of the cell. However, the cytosol is more than just an amorphous fluid: it makes its own contribution to the metabolism of the cell, and is able to interact with certain organelles in order to direct their functions.

1.4.1 The cytosol is the site of several metabolic pathways

We saw earlier (table 1.2) how certain metabolic functions were sequestered within specific organelles; the same is true of the cytosol (table 1.4). In some cases, the enzymes of a particular metabolic sequence in the cytosol may form a multifunctional complex, as does the mammalian fatty acid synthetase, a dimer in which each subunit consists of several covalently linked catalytic polypeptides. Such an association of functionally related proteins will obviously enhance the rate of the overall pathway. Even where there is no obvious physical contact between the participating enzymes of a metabolic pathway, as in glycolysis, the presence of high concentrations of enzymes in the

Table 1.4
Metabolism in the cytosol

Class of metabolite	Metabolic reactions
Carbohydrates	Glycogen synthesis/degradation
	Pentose phosphate pathway
	Glycolysis
	Gluconeogenesis
Fats	Fatty acid synthesis
	Triglyceride synthesis
Amino acids	Synthesis/degradation of many amino acids
Nucleotides	Synthesis of purines and pyrimidines

In some cases, the complete synthesis/degradation of a particular metabolite involves the cooperation of enzymes in the cytosol with those in other subcellular compartments.

cytosol will ensure that the product of one reaction will rapidly become the reactant of the next in the sequence. For example, the cytosolic concentration of glyceraldehyde 3-phosphate dehydrogenase is estimated to be 1.4 mM, twenty times that of its triose phosphate substrate! In other cases, the synthesis or degradation of a particular metabolite may involve the participation of enzymes in the cytosol and in other organelles. For example, the acetyl CoA that is the starting point for the synthesis of fatty acids in the cytosol is obtained from the mitochondria (*via* the breakdown of pyruvate or ketone bodies or ketogenic amino acids).

1.4.2   The cytosol also contains organised structures

The cytosol is not just an aqueous 'soup' within which various metabolic reactions occur. In addition to the multi-enzyme complexes just mentioned, there are various macromolecular structures that make up the *cytoskeleton* of the cell. These include:

(a)   *Microtubules* — hollow cylindrical assemblies (diameter 24 nm) formed by the aggregation of many copies of a **hetero-dimeric** protein, *tubulin*. In animal cells, the microtubules appear to originate from an organising centre in the vicinity of the centriole (section 1.2.1), from where they radiate outwards to the extremities of the cell.

Hetero-dimeric: Having two different subunits.
Homo-dimeric: Having two identical subunits.

(b)   *Microfilaments* — aggregates of actin monomers forming 7 nm diameter filaments. Bundles of microfilaments may run parallel to the long axis of the cell.

(c)   *Intermediate filaments* — a family of protein assemblies (coiled-coil structures roughly 10 nm in diameter) and including the keratins and the neurofilaments, among others. Different classes of intermediate filaments are found in different tissues. Furthermore, the structural proteins of the nuclear envelope (lamins; section 3.2.2) are genetically related to the cytoplasmic intermediate filament proteins.

The organisation of these three types of macromolecular structure in the cytosol can be elegently revealed by immunofluorescence microscopy (section 2.1.1). But what are their functions? Probably microfilaments are involved in cell motility and in the contraction of projections of the cell surface. Microtubules are particularly interesting with regard to organelles. Video-enhanced microscopy studies show membrane-bound vesicles moving down a nerve axon, apparently by interacting in an energy-dependent manner with microtubules. These structures are also thought to be involved in exocytosis, since this process is blocked by *colchicine,* an alkaloid drug that is known to interfere in the assembly of microtubules. Whatever the precise functions of the intermediate filaments, the microfilaments and the microtubules, it is clear that the cytosol contains macromolecular aggregates with organising and contractile potential. The cytosol is more than just an amorphous fluid.

1.5   Principles behind subcellular architecture

In the preceding sections we have been introduced to the major organelles of the cell: the nucleus, the endoplasmic reticulum, the Golgi complex, the lysosome, the peroxisome, the mitochondrion, and the chloroplast. We shall consider each of these organelles in more detail (chapters 3 to 9), after we have familiarised ourselves with the experimental methods that cell biologists apply to the study of their subject (chapter 2). At this point, we should pause to consider what it is that we wish to know about organelles. Are there any common organising principles underlying their structure and function? What can the new techniques of molecular biology tell us about them?

1.5.1   Specific functions imply specific components

It is obvious that each class of organelle is organised so as to carry out efficiently a limited range of subcellular functions (table 1.1). The implications of this statement are two-fold: each organelle must contain a specific set of components (proteins, lipids, carbohydrates, and nucleic acids), and there must be mechanisms for directing newly synthesised components to their appropriate destination within the cell.

Consider for example the lipid composition of subcellular membranes (table 1.5). The data clearly show that each organelle has a membrane with a distinctive set of lipid components. Indeed, some lipids may be unique to a single class of organelle: diphosphatidylglycerol (cardiolipin) in mitochondria, for example. How is this specificity attained? In some cases, the enzymes involved in their synthesis are localised within that

Table 1.5
Lipid composition of subcellular membranes

Lipid class	Plasma membranes	Nuclear membranes	Endoplasmic reticulum membranes		Golgi membranes	Lysosomal membranes	Mitochondrial membranes	
			Rough	Smooth			Inner	Outer
Cholesterol	25	10	6.0	8.5	6.7	14	2.5	4.5
Neutral lipids	16	15	10	8.5	34	17	13	12
Phosphatidylcholine	18	44	55	47	29	24	38	41
Sphingomyelin	13	3.0	3.0	10	7.3	24	2.0	4.0
Phosphatidylethanolamine	11	17	16	18	11	13	21	19
Phosphatidylinositol	3.5	6.5	8.0	6.0	4.9	6.5	5.0	11
Phosphatidylserine	8.0	3.2	3.0	0	2.7	0	1.0	1.5
Diphosphatidylglycerol	0	0	0	0	0	0	15	3.0
Other phospholipids	5.0	1.0	2.0	2.0	5.8	5.0	2.5	3.5

Values are expressed as weight percentages of total lipids, and are averaged data for membrane preparations from mammalian cells. 'Neutral lipids' include cholesteryl esters, tri-, di- and monoglycerides, and free fatty acids. 'Other phospholipids' include phosphatidic acid, phosphatidylglycerol, lysophospholipids, and (in lysosomal membranes) bis(monoacylglycero)-phosphate. Glycolipids may also be present in small amounts, particularly in plasma membranes (up to 5 per cent of the total lipid). The lipid composition of chloroplast membranes is given in table 9.1

organelle; this explanation applies to the galactolipids of the chloroplast, for example (section 9.4.3). In other cases, there is no such obvious explanation. Indeed, the picture is clouded even further by the ability of some membranes to fuse with others, thus leading to mixing of their lipid components. Furthermore, the membrane of the lysosome is derived from that of the Golgi complex, yet their lipid compositions are quite distinct. How is this?

We are faced with the same kind of problem in accounting for the protein composition of the various classes of organelle. One clearly expects to find the enzymes of the Krebs cycle in mitochondria, for example, or the acid hydrolases in lysosomes. How did they get there? One has to assume some kind of signal on each protein that marks it for uptake by a particular type of organelle. The implication of this statement is that each organelle has a specific receptor which recognises that signal. In some cases, these components (signal and receptor) are not necessary, for as we shall see (chapters 8 and 9), mitochondria and chloroplasts have the ability to synthesise certain of their component proteins. However, many more of their proteins are encoded by nuclear genes. How did this particular subcellular distribution of genes come about, and why? Cloning and DNA sequencing techniques should be able to tell us where a particular gene is, how many copies of it there are, and what its structure is. Ideally, one would also want to know what controls the expression of a particular gene. Why are certain enzymes present in lysosomes at concentrations 100-times that of others?

Even when a protein has become localised within a given organelle, it still has to reach the correct final destination: the membrane or the soluble matrix, the inner envelope or the outer envelope. Indeed, the protein must also insert itself with the appropriate orientation: with its functionally active domain exposed at one particular face of a membrane, for example. Furthermore, one type of protein may have to associate with others within a given organelle in order to generate a functionally active complex. All of these specific interactions imply that each protein present in a specific organelle must contain the appropriate structural features, which are in turn determined by the primary amino acid sequence, and ultimately by the base sequence of the corresponding gene. How did that particular gene evolve in that particular way? Does it share any **homology** with functionally related genes?

Homology: A shared structural feature (usually of sequence or higher-order structure). Homology may imply a common evolutionary origin.

1.5.2 Organelles interact dynamically

The preceding section concerned itself mainly with the structure of organelles, but we should not lose sight of their functions, or their origins. In this respect, the chloroplast and the mitochondrion stand out, for they are able to undergo a form of division in order to duplicate themselves. They are semi-autonomous organelles, with their own DNA and protein-synthesising machinery, and with a double membrane around them, rather than the single limiting membrane of most organelles. The components of their biosynthetic apparatus show considerable homology with those of prokaryotes: their circular DNA lacks histones, like that of bacteria; their ribosomal RNA resembles that of bacteria; and their protein synthesis is inhibited by numerous anti-bacterial agents, such as streptomycin. Such is this close similarity that many cell biologists believe that chloroplasts and mitochondria evolved from an association between a prokaryotic organism (a different one for each organelle) and a primordial eukaryote: the so-called **endosymbiont hypothesis** (figure 1.7). Peroxisomes may also have arisen in this way.

Other organelles clearly did not evolve in this way. Lysosomes and secretory granules, for instance, are derived biosynthetically from the Golgi complex. But how do new Golgi complexes form, or new endoplasmic reticulum? In any case, the protein components of all these organelles are synthesised on cytosolic ribosomes by translation of messenger RNA transcribed from the corresponding genes on the

Figure 1.7
The endosymbiont hypothesis of organelle evolution

On at least three occasions in the distant past, prokaryotic organisms are thought to have formed a symbiotic relationship with a primordial eukaryotic cell. Separate evolution of the endosymbionts gave rise to the present-day peroxisomes, mitochondria, and chloroplasts. N, nucleus.

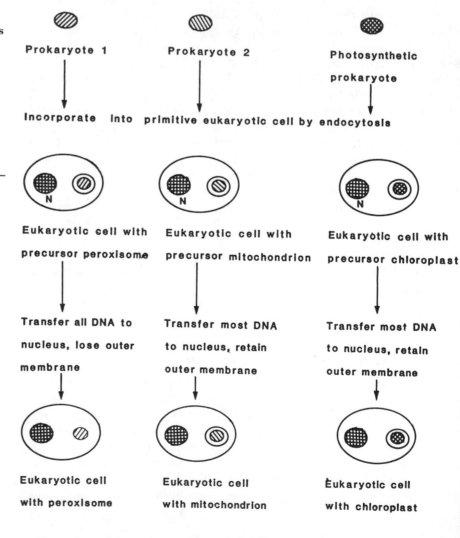

nuclear DNA. In the final analysis, the nucleus directs the biogenesis of all the cell's organelles. Presumably there are also genetic mechanisms for controlling the numbers of each class of organelle in the cell, as well as their precise composition and disposition, factors which in turn should be adapted to the physiological functions of that cell.

In conclusion, there are innumerable structural and functional interactions between subcellular organelles. Although it is advantageous for the cell to localise some of its functions within one particular site, nevertheless the various sites need to communicate with one another, in order to coordinate their metabolism or biological roles. The mechanisms whereby subcellular organelles 'talk' to one another is a major preoccupation of cell biologists, and also of this book.

1.6 Summary

There is enormous variation in the size and shape of eukaryotic (animal and plant) cells. However, they share some common structural features: the nucleus and the surrounding cytoplasm, enclosed within a limiting plasma membrane. Also, they invariably have to carry out certain universal functions: for example, synthesis of RNA in the nucleus; generation of ATP by oxidative phosphorylation in mitochondria; degradation of biopolymers in lysosomes; destruction of potentially harmful oxidising agents in peroxisomes; glycosylation of proteins initiated in the endoplasmic reticulum; packaging of secretory proteins into vesicles by the Golgi complex; and, in photosynthetic cells, the harnessing of light energy by chloroplasts. All these organelles are membrane-bound structures with a characteristic morphology that can be recognised under the electron microscope. The lipid bilayer of their limiting membrane frequently represents a permeability barrier to the free passage of solutes. Membrane-bound proteins act, among other roles, as transporters for these mainly polar solutes; many are asymmetrically arranged so as to facilitate a vectorial (unidirectional) passage of the transported component. The universal organising principles of biological membranes mean that organelles may readily fuse with one another, a feature that is essential to the function of some of them. The organelles are suspended in an aqueous medium, the cytosol, which acts as a common compartment for the exchange of solutes, and contains a highly structured cytoskeleton. New organelles arise either *de novo* by organised assembly of the appropriate components, or by division of existing organelles. Such biogenesis requires the synthesis of specific lipids and proteins, directed invariably by genes in the nuclear DNA. The compartmentalisation of functions within eukaryotic cells has led to extensive dynamic interactions between organelles.

1.7 Study questions

1. Name the major subcellular organelles, with one representative

function for each, of animal cells. Which additional organelle is found in some plant cells?

2. What proportion of the total volume of a yeast cell (diameter 10 μm) is occupied by the nucleus (diameter 2.4 μm)? (Volume = $4/3\ \pi r^3$)

3. Acetyl CoA cannot cross the inner mitochondrial membrane, which lacks a transport protein for it. How then does the mitochondrion provide the cytosol with acetyl CoA for fatty acid synthesis?

4. The peroxisome contains the enzyme catalase, which removes potentially harmful hydrogen peroxide according to the reaction: $2H_2O_2 \rightarrow 2H_2O + O_2$. What other enzymes is the peroxisome likely to contain?

5. The lysosome fuses with other vesicles in order to carry out the degradation of biological macromolecules, catalysed by hydrolase enzymes active at acid pH (pH 5 or below). The surrounding cytosol is at pH 7.4. Predict which proteins you might expect to find in the lysosomal membrane.

6. How could conventional electron microscopy be adapted in order to confirm that lysosomes contain acid hydrolase enzymes?

1.8   Further reading

The ultrastructure of the cell is adequately described in any of the textbooks mentioned in the Introduction.

Reviews and books

de Duve, C. (1985). *A Guided Tour of the Living Cell,* Scientific American Books, W. H. Freeman and Co., New York and Oxford.
(A highly personal and anthropomorphic account)
Trends Biochem. Sci., (1986). **11**, 437–483.
(Has detailed reviews of various aspects of cell organisation)

Structure and evolution of cellular components

Weinberg, R. A. (1985). *Sci. Amer.,* **253**(4), 34–43.
(Introduction to the molecules of life)
Bretscher, M. S. (1985). *Sci. Amer.,* **253**(4), 86–90.
(Introduction to cell membranes)
Weber, K. and Osborn, M. (1985). *Sci. Amer.,* **253**(4), 92–102.
(Introduction to the cytosol)
Srere, P. A. (1987). *Ann. Rev. Biochem.,* **56**, 89–124.
(Organised assemblies of metabolic enzymes)
Margulis, L. (1971). *Sci. Amer.,* **225**(2), 48–57.
(The endosymbiont hypothesis)

Subcellular organelles

Detailed accounts of the structure, function and biogenesis of the major organelles of animal and plant cells are presented in chapters 3 to 9.

2 THE EXPERIMENTAL STUDY OF ORGANELLES

Cell biologists use a variety of techniques to study the structure and function of organelles. These experimental approaches can be broadly classified into imaging methods, biochemical (including immuno-chemical) methods, and recombinant DNA methods. Obviously there is not the space here to describe all these techniques in detail (see section 2.6, Further reading, for more comprehensive treatments). However, we need to look briefly at some of them, in order to understand how cell biologists set out to answer a particular question, and what information can be derived from the techniques available to them. We will make frequent references throughout this book to the experimental methods described in this chapter, which is best used as a resource to be consulted, rather than to be read from beginning to end.

2.1 Imaging methods

The ability to generate an image depends on the absorption or scattering of light (or other electromagnetic radiation) by an object, and a system for detecting the resulting change. This principle underlies the various forms of *microscopy* and *diffraction methods*. A key feature of these techniques is their *resolution* or resolving power: the shorter the wavelength of the incident radiation, the greater the potential resolution. An important consideration in applying these methods to cells is that the resulting image is very faint unless the components of the cell are first *fixed* and *stained* in order to improve the contrast. Stains are often non-specific, but they may show some specificity for a particular subcellular component. Selective staining may also be possible if one can exploit a specific property of the component concerned: an antigenic site on a structural protein, for example, or the active site of an enzyme. Although imaging methods are mainly qualitative, they may in some cases be adapted to provide quantitative information also. In recent years computers have been increasingly exploited in order to enhance images, as well as to generate images from data obtained by other means. Overall, however, we do not have much information on the three-dimensional structure of most organelles and their components. Furthermore, the images we have are invariably static; there are no widely applicable techniques that visualise the dynamic functions and interactions of organelles within the cell.

2.1.1   Light microscopy

We saw earlier (section 1.1.2) how light microscopy provided the first pictures of the cell and its more obvious components. The procedure adopted in a typical application of this technique is given in figure 2.1. One of the criticisms levelled at this approach is that the organic chemicals used for fixation may introduce artefacts. These are less likely if one uses *frozen specimens*, and a cooled microtome for cutting sections; even here, a method for staining is required.

Are the features observed by such light microscopy also evident in living cells (unfixed and unstained)? The answer, in the affirmative, has been provided by modified forms of light microscopy that allow cell biologists to observe cells directly: *phase-contrast, Normarski differential-interference contrast,* and *dark-field* microscopy. A further advantage of such techniques is that they allow time-lapse photographs to be taken, so that changes in cell structure or function can be observed over extended periods of time.

The specificity of the stains commonly used in light microscopy is rather limited: haematoxylin preferentially stains nucleic acids, for example, but other dyes show up all subcellular components. Specificity can be enhanced by several modifications. Protease treatment prior to staining with Giemsa reveals the characteristic banding patterns of mitotic chromosomes, such that the **karyotype** of a cell can be determined (figure 3.1). If one is interested in the tissue distribution of an enzyme (*histochemistry*), the section can be incubated with a substrate which the enzyme can convert to a coloured, insoluble product; the site of deposition of the product marks the location of the enzyme. Alternatively, one may exploit the antigenic properties of a protein of interest. Incubation of the section with a specific antiserum will result in the protein binding the antibodies; these can be detected with a second antibody coupled to a fluorescent dye. This approach

Karyotype: The number and morphology of the chromosomes in a cell.

Figure 2.1
Experimental procedures in light microscopy

The approach depicted here generates images of non-living cells. Individual components of cells may be highlighted by selective staining (for example, the haematoxylin stain for DNA and RNA). Light microscopy of living cells entails a different approach (for instance, the use of phase-contrast or Nomarski optics) which omits the fixation, sectioning, and staining procedures.

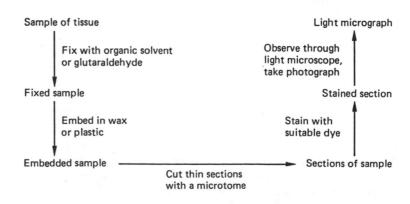

is called **immunofluorescence microscopy.** (Alternatively, the second antibody may be tagged with an enzyme such as peroxidase.) One can exploit fluorescence in other ways, too. For example, certain quinoid drugs, such as mepacrine, fluoresce only when exposed to acidic pH; incubation of cells with such compounds illuminates acidic compartments within the cell, such as lysosomes. Also chemicals such as quin-2, which fluoresce only in the presence of calcium ions, can reveal subcellular sites of relatively high Ca^{2+} concentration.

In general terms, however, light microscopy is of rather limited value in the study of subcellular architecture. Its limit of resolution ($0.2\ \mu$m) is barely adequate to discern a mitochondrion (diameter, $0.5\ \mu$m). We need an imaging technique of much greater resolving power.

2.1.2 Electron microscopy

The principle underlying *transmission electron microscopy* (TEM) is essentially the same as that of conventional light microscopy: components of the sample scatter some of the incident radiation, creating the contrast that gives rise to an image. Here, however, the radiation is a beam of electrons, accelerated with a very high voltage (10^5 volts or more) to produce an extremely short wavelength (0.2 nm or so). The result is an imaging technique with a very high resolution; even with the practical limitations inherent in the method, a resolving power of 2 nm can be achieved.

Because biological specimens contain elements of low atomic number, they do not readily scatter electrons, a process that requires the presence of heavy metal atoms. Accordingly, fixing of the sample usually involves treatment with osmium tetroxide, which cross-links both proteins and lipids, as well as with glutaraldehyde. The fixed specimen is stained with a solution of heavy metal salt, such as lead acetate or uranyl acetate. The preparation of the very thin sections (50–100 nm) requires a special ultramicrotome. We have already seen the results of TEM of an animal cell (figure 1.3) and a plant cell (figure 1.4).

TEM can be modified to provide enhanced specificity, in much the same way as light microscopy can. Histochemistry can be performed at the EM level if the product of the enzyme-catalysed reaction is electron-dense. Specific components can be visualised by immuno-chemical staining if the second antibody is coupled to an electron-dense marker, such as ferritin (an iron-containing protein). Increasingly, immuno-EM is being performed with antibodies labelled with complexes of gold or silver salts (figure 2.2). These complexes can be prepared so that they have a characteristic and constant diameter. Thus, two different antibodies can be tagged with complexes of different sizes, so that a section under TEM can be made to reveal two components simultaneously, since one will be associated with relatively large electron-dense spots (say 9 nm) and the other with relatively small spots (6 nm).

Fluorescence: Some compounds absorb light of one wavelength, and emit light of a longer wavelength. The emitted light can be detected with a fluorimeter or a fluorescence microscope. (An example of immuno-fluorescence microscopy is depicted in figure 6.7.)

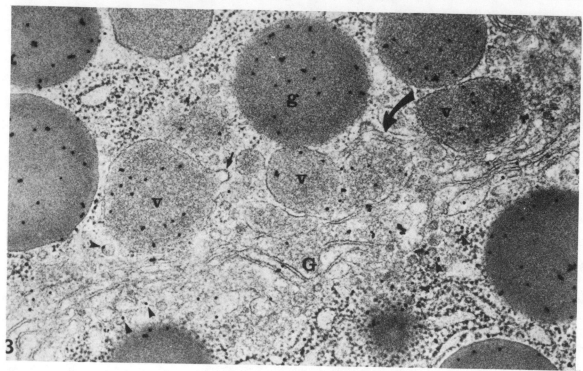

Figure 2.2
Immuno-electron microscopy

This thin section of acinar cells from
rat pancreas was labelled for amylase
by the Protein A–gold technique.
Here, an appropriate series of
incubations results in amylase binding
specific antibody, which is
subsequently recognised by Protein A
labelled with colloidal gold. The
electron-dense (black) gold particles
are present over the rough ER
(arrowheads), Golgi complex (G),
condensing vacuoles (v), and
zymogen granules (g). × 54 000.
[Taken from Bendayan *et al.* (1986).
Amer. J. Anat., **175**, 379–400;
reproduced by permission of the
publisher, Alan R. Liss Inc.]

TEM provides a two-dimensional cross-section through a specimen
at only one point, so the interpretation of such images in three
dimensions cannot be made with confidence. There are several
approaches that overcome this problem. One involves making serial
sections of adjacent points through the cell, and reconstructing the
objects observed in the TEM images. Another is to use *scanning electron
microscopy* (*SEM*). Here, the surface of the object is *shadowed* with a
thin layer of heavy metal, and the image obtained from the electrons
scattered at right angles to the incident beam. The resolution of SEM,
10 nm or so, is less than that of TEM, but it does provide detailed
pictures of the surfaces of cells (figure 2.3). A three-dimensional effect
of the intracellular contents can also be recreated by using *field-emission
SEM* and suitably treated sections (figure 2.4).

Two further related techniques can provide three-dimensional
information on subcellular architecture. In *freeze-fracture EM* the frozen
specimen is sheared with a glass or diamond knife such that the fracture
plane may pass through the centre of the lipid bilayer of the membrane.
Intramembranous proteins are thereby exposed. The surface of the
specimen is shadowed with platinum and carbon, and the specimen
dissolved away with strong alkali, before viewing the surface replica
by TEM. In the related *freeze-etch EM* the fractured specimen is
subjected to a vacuum in order to remove some of the ice (by
sublimation) and thereby further expose the surfaces of the cell's
contents (figure 2.5).

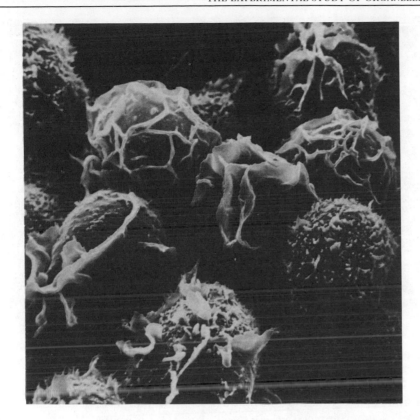

Figure 2.3
Scanning electron microscopy

This type of EM generates dramatic three-dimensional images of the cell surface. These leukocytes (white blood cells) have been stimulated by antigen; their plasma membrane has numerous short projections (microvilli) and more extensive 'ruffles'. × 4000. [Taken from Phillips, D. M., Bashkin, P. and Pecht, I. (1985). *J. Ultrastruct. Res.*, **90**, 105–113; reproduced by permission of the publisher, Academic Press Inc.]

Finally, individual macromolecules such as proteins and nucleic acids, or supramolecular assemblies such as ribosomes and viruses, can be viewed under the transmission electron microscope by *negative staining*. In this technique, salts of heavy metals, like phosphotungstate or phosphomolybdate, are used to create a dense background relative to the outline of the specimen. This method is especially good for observing fine structure at membrane boundaries. Also, TEM can be coupled to *autoradiography* in order to reveal the subcellular location of a particular component labelled with a radioactive isotope.

2.1.3 Diffraction methods

Where an object consists of arrays of regularly spaced components, then diffraction methods can be applied to create an image. The best example is *X-ray crystallography*, where the diffraction of X-rays (wavelength 0.1 nm) by the atoms in a crystal allows the reconstruction of three-dimensional images at the submolecular level. There are practical problems associated with this technique as applied to biological macromolecules such as proteins: it is not always easy to obtain suitable crystals, and one has to overcome the 'phase problem' inherent in the method, usually by producing a derivative of the protein substituted with a heavy metal. The mass of diffraction data generated can now be readily dealt with by powerful computers in order to

Figure 2.4
Scanning EM of organelles

The interior of cells can be examined by scanning EM, provided that an appropriate fixing technique is used, and that most of the cytoplasmic proteins are first removed. This micrograph provides a spectacular three-dimensional view of the intracellular structures of a nerve cell from rat spinal cord. Clearly visible are mitochondria (M), Golgi complex (G), rough ER (rER), smooth ER (sER), and apparent connections between Golgi complex and smooth ER (arrows). [Courtesy of K. Tanaka.]

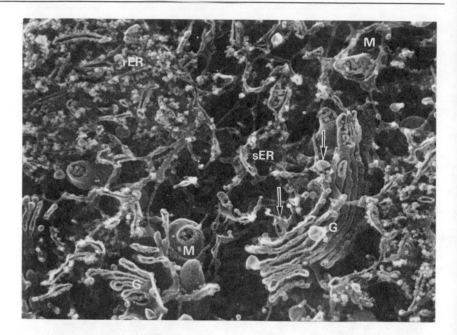

generate a model of the macromolecule at the atomic level. Many soluble proteins, nucleic acids and other macromolecules have been examined by X-ray crystallography. In the future, it should also be possible to use modified techniques to view membrane proteins and complex macromolecular assemblies such as ribosomes.

Diffraction methods can also be applied at the light level, by using laser beams of defined wavelength, so as to enhance images obtained by electron microscopy or other means. Computer programmes are

Figure 2.5
Freeze-etch electron microscopy

This micrograph shows the apical cytoplasm of an epithelial cell from guinea-pig trachea. Cilia, with their characteristic 9 + 2 arrangement of microtubules, can be seen emerging from basal bodies. × 55 000. [Taken from Arima, T., Shibata, Y. and Yamamoto, T. (1984). *J. Ultrastruc. Res.*, **89**, 34–41; reproduced by permission of the publisher, Academic Press Inc.]

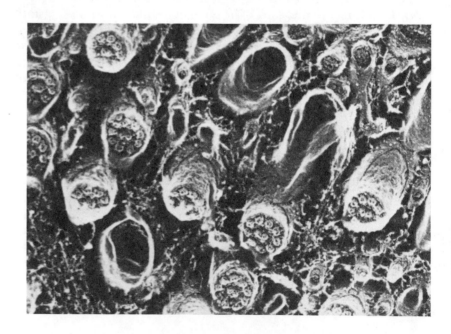

available that will remove unnecessary detail and reconstruct the essential features of the observed object.

2.2 Biochemical and immunochemical methods

The morphology of subcellular components revealed by imaging techniques (section 2.1) needs to be supplemented by further information on their structure and function. This is where biochemical and immunochemical methods are so useful. Ideally, one might like to prepare reasonable amounts of a single class of organelle, in order to study the composition and function of its individual components, and how they interact. One must always bear in mind that such an approach may generate artefacts, or may only provide information on one organelle without describing its relationships with other cellular components. We obviously cannot cover here all the biochemical and immunochemical methods used by cell biologists; we will consider only the most important.

2.2.1 Centrifugation

Reasonably pure preparations of each class of organelle can be obtained by separating the components of an isotonic tissue homogenate on the basis of their behaviour in the *ultracentrifuge*. Such machines generate centrifugal forces of $10^5 \cdot g$ (100 000 times the force of gravity) or more, under which conditions a particle will **sediment** at a rate dependent on a number of factors: its morphology and density are particularly important. Particles may be fractionated either by *differential centrifugation* (on the basis of sedimentation rate) or by *isopycnic* (*density-gradient*) *centrifugation* (on the basis of buoyant density).

Sediment: Move towards the bottom of the container, usually (in centrifugation studies) a plastic tube.

A typical example of a scheme for the separation of the organelles of mammalian liver by differential centrifugation is given in figure 2.6. The salient features of this method are:

(a)   The cells must be gently disrupted in an isotonic buffered medium in order to release their contents. Small pieces of tissue, rapidly isolated from the experimental animal, may be *homogenised* by a Teflon pestle rotating inside a tight-fitting cylindrical glass mortar (the Potter–Elvejhem *homogeniser*), but other methods of homogenising are available. This process may be subject to artefacts: for example, the endoplasmic reticulum is invariably broken up into small vesicles called *microsomes*.

(b)   The homogenate is subjected to increasing centrifugal forces, and at each step a pellet (of sedimented particles) is separated from a supernatant, which is centrifuged further.

(c)   The final supernatant represents the cytosol of the cell.

(d)   The organelles present in each pellet must be identified. Routinely, this is achieved either by electron microscopy or by the assay of

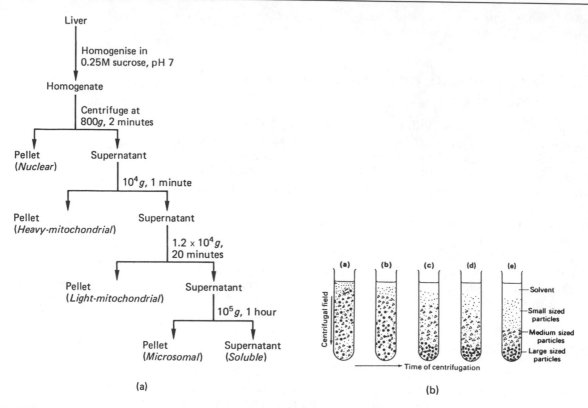

(a)   (b)

Figure 2.6
Differential centrifugation

(a) An isotonic homogenate of mammalian liver is subjected to centrifugation at increasing (force × time) integrals. In practice, the pellets are often re-suspended and re-centrifuged. (For an example of representative results, see figure 6.2.) (b) At each centrifugation step, the mixture of particles, originally uniformly distributed [tube (a)], undergoes sedimentation such that the larger particles reach the bottom of the tube before the smaller particles [tubes (b)–(e)]. Note that the pellet of large-sized particles [tube (e)] is contaminated with smaller particles. [Taken from *Principles and Techniques in Practical Biochemistry* (Wilson, K. and Goulding, K. H., eds), 3rd edn, Edward Arnold, London; reproduced by permission of the publisher.]

marker enzymes (or other components). These enzymes are known from other studies to be present in only a single class of organelle; for example, acid phosphatase is localised in the lysosomes (section 6.1.1). The assay medium usually contains a detergent to break the membrane of the organelle, since otherwise the membrane-enclosed marker enzyme would not gain access to the substrate in the medium (so-called *structure-linked latency*).

However, the value of differential centrifugation is limited by the difficulty in separating organelles with similar sedimentation properties, in particular mitochondria, lysosomes, and peroxisomes. Such particles can be further fractionated by density-gradient centrifugation. In particular, peroxisomes (density = 1.09 g cm^{-3} in 0.25 M sucrose) can be separated from lysosomes and mitochondria (density about 1.10 g cm^{-3}). These two latter organelles can be resolved if the experimental animal is first treated with Triton WR1339, which accumulates in lysosomes and reduces their density. Various media are available for generating density gradients: concentrated sucrose solutions (for separating organelles), or concentrated caesium chloride solutions (for separating nucleic acids), for example. Percoll, a polymer-coated colloidal silica preparation, generates a density gradient as the centrifugation proceeds, and can be used to separate cells as well as organelles.

Any centrifugation method has inherent problems: the variable morphology of organelles, for example, or artefacts introduced by homogenisation. However, despite these drawbacks, centrifugation techniques have provided cell biologists with a powerful means for obtaining purified preparations of organelles for further study.

2.2.2 Cell culture

Intact cells can be maintained in a viable state by various methods of *cell culture*. Isolated cells may be obtained in their natural state (blood leukocytes, for example) or by dissociation from a tissue sample (such as liver hepatocytes or skin fibroblasts); in some cases, aggregates of cells are used (such as pancreatic islets of Langerhans). The composition of the culture medium is important for the long-term survival of the cells. It invariably contains glucose, salts, buffers, amino acids, vitamins and other essential nutrients, as well as an additive of ill-defined composition, such as foetal calf serum; anti-microbial agents are frequently included to limit the risk of infection. It is important to control the pH and temperature of the medium, and the composition of the atmosphere above the medium (where the medium is buffered with bicarbonate). Cells may grow free in suspension or attached to polysaccharide beads, or they may attach to a fixed substratum (the bottom of a plastic culture flask, for instance), where they invariably form a *monolayer* of adjacent cells. The anchored cells can be released by brief exposure to the proteolytic enzyme, trypsin, or by scraping the surface with a rubber-tipped rod.

The ability to culture cells *in vitro* provides a means of preparation of reasonable amounts of a defined population of cells for further study. For example, the cells' organelles can be analysed by centrifugation (section 2.2.1), or the biosynthesis of a given protein can be followed by pulse-labelling (section 2.2.5b).

In vitro: Literally 'in glass'. An expression used to indicate an experiment done under controlled artificial conditions in the laboratory. *In vivo*: Literally 'in life'. Applied to a process as it occurs in the living organism.

2.2.3 Electrophoresis

Numerous biochemical techniques rely upon the migration of a charged macromolecule in an applied electrical field. Such separations are usually carried out in a gel medium prepared in a buffer of suitable pH. After *electrophoresis* the separated components have to be identified with a suitable stain; this may be relatively non-specific (for example, Coomassie Blue for proteins) or highly specific (detecting the product of an enzyme-catalysed reaction, for instance). This general approach is summarised in figure 2.7.

There are many electrophoretic methods. Most resolve soluble components on the basis of their charge. Thus, a protein exposed to a pH more alkaline than its **isoelectric point**, pI, will be negatively charged and will migrate towards the anode (positive electrode); if the pH is on the acidic site of the pI, then the protein migrates to the cathode. Numerous supporting media have been used for the electrophoresis of proteins, including cellulose acetate, starch, agarose, and polyacrylamide, some in the form of gradient gels of increasing polymer concentration. Numerous buffer systems have been described, also.

Isoelectric point: The pH at which a protein in solution bears an overall zero charge.

Figure 2.7
Separation of proteins by electrophoresis

Polyacrylamide gel in Tris-HCl buffer, pH 8.9, was prepared at a concentration of 5 per cent (w/v) (stacking gel) or 10 per cent (w/v) (separating gel). Sample wells A, B, and C contained proteins with isoelectric points of 8.0, 6.0, and 4.0 respectively; D was a mixture of all three. The top of the stacking gel and the bottom of the separating gel were immersed in electrode buffer (Tris-borate, pH 8.8). After electrophoresis (anode at the bottom), the separated proteins were 'fixed' by denaturing them with trichloroacetic acid solution (TCA) prior to staining. Alternatively, the gel could have been stained (without fixing) to detect the biological activity of individual proteins (such as enzyme activity).

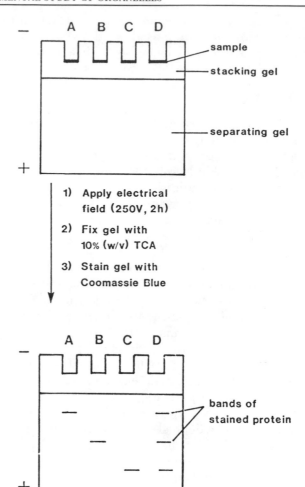

$CH_3(CH_2)_{11}OSO_3^- Na^+$: Structure of SDS, sodium dodecyl sulphate.

After electrophoresis, proteins may be 'fixed' by soaking the gel in a denaturing solution (such as 10 per cent trichloroacetic acid), prior to staining. Alternatively, if the biological activity of the protein is to be retained, the gel is immediately incubated with the detection reagent and observed shortly afterwards.

Separation on the basis of charge also underlies *isoelectric focusing*, where proteins migrate in a pH gradient, under the influence of an applied electrical field, until they come to rest at their isoelectric point. This method has a higher resolving power than that of electrophoresis, but is technically more difficult to carry out.

Some electrophoretic techniques separate components on the basis of molecular size. Proteins treated with the detergent **sodium dodecyl sulphate** (**SDS**) become denatured and coated with a uniform negative charge. At pH 8.3, they all migrate in a polyacrylamide gel to the anode, but at different rates depending on their molecular mass. Thus, the smallest migrate most rapidly towards the end of the gel (figure 2.8). The great virtue of this technique (*SDS-PAGE*) is that it

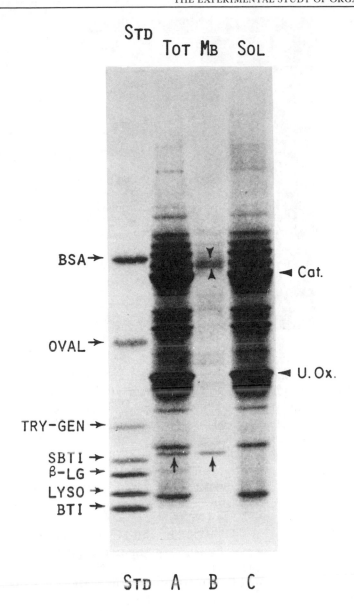

STD

TOT MB SOL

BSA →

OVAL →

TRY-GEN →

SBTI →
β-LG →
LYSO →
BTI →

◄ Cat.

◄ U. Ox.

STD   A   B   C

Figure 2.8
SDS-PAGE separation of polypeptides

In the presence of sodium dodecyl sulphate (SDS), denatured polypeptides migrate on electrophoresis at a rate inversely proportional to their molecular mass. This is demonstrated in the lane (STD) which contains marker proteins of known size (in parentheses): BSA, bovine serum albumin (68 kDa); OVAL, ovalbumin (45 kDa); TRY-GEN, trypsinogen (24 kDa); SBTI, soybean trypsin inhibitor (21.5 kDa); β-LG, β-lactoglobulin (18.4 kDa); LYSO, lysozyme (14.3 kDa); and BTI, bovine trypsin inhibitor (6.5 kDa). The remaining lanes contain protein (100 μg) of peroxisomes isolated from rat liver. TOT, total proteins; MB, membrane-bound proteins; SOL, soluble proteins (including catalase, Cat., and urate oxidase, U. Ox.). [Taken from Fujiki, Y. *et al.* (1982). *J. Cell Biol.*, **93**, 103–110; reproduced by copyright permission of the Rockefeller University Press.]

can be applied not only to soluble proteins, but also to membrane-bound proteins, since the SDS extracts such proteins from the lipid bilayer. Furthermore, the migration of a given protein provides a measure of its molecular mass (in kiloDaltons, kDa), provided that the gel is calibrated with standard proteins of known size. Glycoproteins behave abnormally in this system however; they migrate as components of apparently larger size.

Nucleic acids bear a uniform negative charge (from the phosphate groups of the sugar–phosphate backbone), and they migrate in an electrical field towards the anode at a rate that depends on their molecular mass. As with proteins in SDS-PAGE, the smaller components

Figure 2.9
Separation of nucleic acids by electrophoresis

Large fragments (> 50 kb) of DNA can only be resolved on the basis of size by pulsed-field gel electrophoresis, as shown here. Samples were DNA from a human female (lane 1) and from a male with a large deletion from the X chromosome (lane 2). DNA was digested with the restriction endonuclease *Sfi*I prior to electrophoresis (anode at the bottom). A, gel pattern of DNA bands stained with ethidium bromide. The outer lanes contained size markers, which appear as discrete, light bands (but note the curved distortion). B, Southern blot of the gel pattern, hybridised with a radioactive probe which recognises the gene for the X-linked enzyme, glycerol kinase. A fragment of 800 kb appears as a dark band in the control lane, but it is missing in the deleted DNA lane. [Taken from van Ommen, G. J. B. and Verkerk, J. M. H. (1986), in *Human Genetic Diseases – a Practical Approach*, (Davies, K. E., ed.), IRL Press Ltd, Oxford; reproduced by permission of the publisher.]

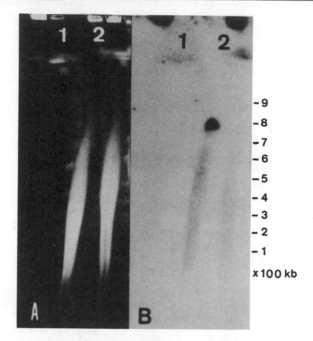

Intercalate: Insert between the base pairs of DNA.

DEAE: Diethylaminoethyl, $- CH_2CH_2N(C_2H_5)_2$.

migrate further than the large ones (figure 2.9). Once again, size calibration of the gel (in this case, an agarose gel) is possible with DNA fragments of known size (such as phage λ DNA digested with the restriction endonuclease *Hin*dIII, to give fragments of 23.1, 9.42, 6.56, 4.36, 2.32, and 2.03 kb). The resolved bands of nucleic acid are usually detected with *ethidium bromide*, which **intercalates** into double-stranded nucleic acid and can be detected by its orange fluorescence when illuminated with ultra-violet light. Thus, DNA is detected more readily than RNA, and large fragments better than small fragments. The useful fractionation range for this form of electrophoresis is between 1 kb and 10 kb. Larger fragments (50–1000 kb) can be separated by *pulsed-field electrophoresis* in agarose gel (figure 2.9), smaller fragments by electrophoresis in polyacrylamide gel.

The various electrophoretic methods described above can obviously be coupled to other analytical techniques: specific proteins can be detected by antibodies (section 2.2.7) for example, or specific DNA or RNA fragments by hybridisation with radioactive probes (section 2.3 and figure 2.9).

2.2.4   Chromatography

The separation of macromolecules by *chromatography* exploits certain structural or functional properties of the components concerned. An incomplete list of the techniques available would include the following:

(a)   *Ion-exchange chromatography*: separates proteins on the basis of charge. Negatively charged proteins will bind to a positively charged matrix (such as **DEAE-cellulose**) at an appropriate pH. Bound proteins can be displaced (*eluted*) by increasing the ionic strength of the solution passing over the matrix, or by decreasing the pH until it is more acidic than the pI of the protein of interest.

(b)   *Chromatofocusing*: separates proteins on the basis of isoelectric point. The principle is the same as that behind isoelectric focusing (section 2.2.3), but the practical procedure resembles that for ion-exchange chromatography.

(c)   *Gel filtration*: separates macromolecules on the basis of size. A beaded matrix with pores of defined size allows smaller molecules to penetrate, but not larger ones. The largest proteins/nucleic acids thus emerge first from the gel matrix. This method can give information on molecular mass (if the gel column is calibrated with markers of known size), or be used to separate macromolecules from low-molecular-mass contaminants (desalting).

(d)   *Hydrophobic interaction chromatography*: depends on the binding of exposed non-polar groups of a protein to a non-polar compound immobilised on a gel matrix. Bound proteins are eluted by disrupting the hydrophobic interactions (with low-ionic-strength solutions, or with ethylene glycol).

(e)   *Affinity chromatography*: depends on the specific binding of a macromolecule to an appropriate *ligand,* covalently coupled to a gel matrix. Bound components may be eluted with a solution of a second ligand, which competes for the binding site on the macromolecule. Ligands may be relatively small (a competitive inhibitor might be used for purifying an enzyme) or relatively large (gel-coupled insulin for purifying the insulin receptor). Most species of eukaryotic messenger RNA can be purified by affinity chromatography on a matrix substituted with oligo-deoxythymidine (oligo(dT), with about 15 residues of deoxy-thymidylate), or with polyuridine (poly(U), with about 100 residues of uridylate). The rationale behind these approaches is that the mRNA has a long tract of adenylate residues (*poly(A) 'tail'*) at its 3' end, and there is complementary base pairing between the poly(A) and the oligo(dT) (or poly(U)). Bound mRNA is eluted by disrupting the base pairing, with a solution of low ionic strength or one containing formamide. This form of affinity chromatography is an essential early step in the preparation of complementary DNA for cloning purposes (section 2.3).

(f)   *High-pressure liquid chromatography (h.p.l.c.)*: all the above techniques can also be performed at very high pressures, by using columns packed with a suitable non-compressible matrix, and high-pressure pumps. The main advantage here is a dramatic reduction in the time required for the chromatography, with a consequent reduced risk of denaturation of labile components.

(g)   *Tests of purity*: when one has purified the desired component, one obviously wants to check if it is pure (that is, free from all contaminating macromolecules with similar properties). Hence, one needs a technique that combines high resolving power with a detection method of great sensitivity; thereby even tiny amounts of a contaminant with almost identical physico-chemical properties can be detected. For proteins, isoelectric focusing in polyacrylamide gel and SDS-page are frequently employed; minute amounts (in the ng range) of extraneous protein can be visualised with a highly sensitive *silver stain*.

2.2.5   Biosynthesis

Cell biologists frequently need to study the *biosynthesis* of a desired component, particularly proteins and nucleic acids. Two common approaches are as follows:

(a)   In vitro *translation*: mRNA (purified by affinity chromatography; section 2.2.4e) can be used to direct the synthesis of the corresponding polypeptides *in vitro*. What is required is a suitable translation system containing ribosomes, GTP, accessory proteins and enzymes, amino acids, transfer RNA, and ATP, but lacking endogenous mRNA. Such complex mixtures can be prepared from lysates of wheat germ or of rabbit reticulocytes (red cell precursors). Usually one amino acid of the twenty is radioactive, so that protein biosynthesis can be followed by measuring the rate of incorporation of radioactivity into material precipitated by trichloroacetic acid (that is, protein). Radioactive polypeptides can be separated by SDS-PAGE, and detected in the dried gel by **autoradiography** or **fluorography**.

Autoradiography and **fluorography**:   Radioactive isotopes, such as ^{3}H, ^{14}C, ^{35}S and ^{32}P, emit radiation (principally β-rays) which will cause the reduction of silver ions in a photographic emulsion, and thereby form a blackened image after the film is developed. This process may be direct, as in autoradiography, or indirect, as in fluorography. In the latter case, the gel is impregnated with a compound that fluoresces when exposed to radiation, and the photographic image is created by the emitted fluorescence.

(b)   *Pulse-labelling*: the biosynthesis of a particular component *in vivo* can be followed by culturing cells in a medium containing a suitable radioactive precursor (^{35}S-labelled methionine for a protein, for example, or ^{32}P-labelled ATP for RNA). After a suitable incubation time (the '*pulse*'), the radioactively labelled components are extracted and analysed. An example of such a study applied to a protein is shown in figure 2.10. In some experiments the pulse-labelling may be followed by a '*chase*', with continued incubation of the cells in non-radioactive ('cold') medium. The further processing of labelled components can thereby be followed (figure 2.10). Post-translational processing (after the protein is released from the ribosome) may also be followed by incubating radioactive protein with microsomes prepared from dog pancreas. The microsomes mimic the endoplasmic reticulum (ER), and contain the components required for transporting secretory proteins across the ER membrane and the subsequent covalent attachment of sugar residues (glycosylation).

2.2.6   Membrane transport

The passage of a solute across an intracellular membrane can be followed in several ways, both direct and indirect. A prerequisite

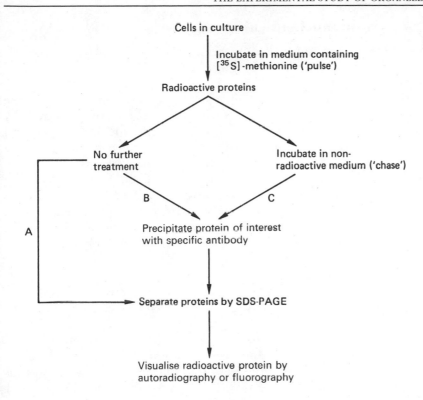

Figure 2.10
Pulse-chase studies of protein biosynthesis

The total labelled proteins can be analysed (A). Alternatively, an individual protein can be singled out for analysis, either immediately after the radioactive 'pulse' (B), or later following a 'chase' period (C).

for any method is a suitable membrane preparation, as obtained by centrifugation of an isotonic tissue homogenate, for example (section 2.2.1). The rate of solute transfer is frequently used as a measure of the activity of the transport protein involved, where facilitated diffusion or active transport is concerned; passive diffusion does not require such a mediator (table 1.3).

Radioactive labelling can be used to trace the path of some solutes. These may be ions (such as $^{22}Na^+$) or macromolecules tagged with a labelled component. The transfer of the solute from one compartment to another across the membrane can be followed by counting the radioactivity in the two compartments — by *liquid scintillation counting* (for β-emitters) or by *gamma counting* (for γ-emitters).

In the case of proteins, their transfer across the membrane can be assessed indirectly by adding a protease solution to the initial compartment at timed intervals. Any protein that has not crossed the membrane before the addition will be proteolytically degraded, whereas any transported protein will be inaccessible to the enzyme, which cannot itself cross the membrane.

The study of transport proteins frequently involves their purification in the presence of detergent, followed by the incorporation of the purified protein into an artificial lipid bilayer (either as a membrane sheet, or as membrane vesicles called *liposomes*). Such *reconstitution experiments* allow the researcher to investigate the effect of external conditions (pH, temperature, etc.) on the activity of the membrane transport protein, or the effect of the lipid environment itself.

2.2.7   Immunochemical methods

Immunoglobulins (antibodies) have long been exploited by biochemists by virtue of their ability to recognise specific sites (epitopes) on their corresponding antigen. We saw earlier (section 2.1.1) how microscopists could use antibodies as labelling reagents in immunofluorescence microscopy. Up until recently such antibodies were prepared by injecting the component of interest into an experimental animal, and waiting for its immune response to generate an *antiserum*. Such antisera were not specific for a single epitope, however, and were said to be *polyclonal*. It is now possible, thanks to work on **hybridomas** by **Milstein** and **Kohler**, to prepare *monoclonal antibodies*, which bind to a single epitope with a constant affinity. Since most immunoglobulins are divalent (they each contain two antigen-binding sites), they can form a lattice-like *immune complex* when they react with an antigen that has several epitopes on the same molecule. Another significant property of antibodies is their high degree of specificity. If an antibody *cross-reacts* with two different proteins, for example, then these proteins almost certainly share some common structural elements, and hence they may also be related genetically.

We do not have space to consider in detail the many types of immunochemical method. Many rely on the interaction of antibodies with their corresponding antigens in solution, usually in a supporting medium such as agarose gel. Thus, we have immunodiffusion and various forms of immunoelectrophoresis. Others rely upon the interaction of antibodies with antigens on the surfaces of cells. In each case, the specificity of the immunoglobulin allows one to analyse individual components of a complex mixture. One technique that has been widely applied in the study of proteins is *Western blotting* (figure 2.11). The term 'blotting' refers to the transfer of the resolved protein components from the polyacrylamide gel onto a nitrocellulose membrane. Since

Hybridomas: Immortal antibody-producing cells, produced by fusing B-lymphocytes exposed to antigen with myeloma cells (cancerous B-lymphocytes).

Figure 2.11
Western blotting of proteins

This technique allows the detection of a single protein in a mixture subjected to SDS-PAGE. As shown here, the detector antibody is radioactively labelled (indicated by the asterisk). More commonly the detector antibody is covalently coupled to the enzyme, peroxidase. Subsequent incubation with appropriate substrates (such as 3,3'-diaminobenzidine and H_2O_2) generates an insoluble brown product at the site of the protein. (Note that *all* the proteins are transferred from the gel to the nitrocellulose filter, not just the one detected here.) [Taken from Macleod, A. and Sikora, K. (1984). *Molecular Biology and Human Disease*, Blackwell Scientific Publications, Oxford; reproduced by permission of the publisher.]

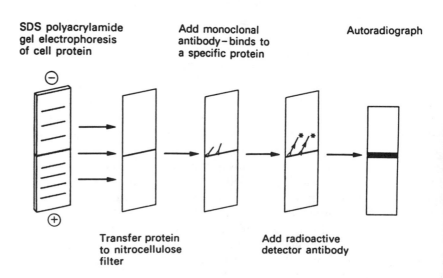

SDS polyacrylamide gel electrophoresis of cell protein

Add monoclonal antibody – binds to a specific protein

Autoradiograph

Transfer protein to nitrocellulose filter

Add radioactive detector antibody

the final step in the procedure is an enzyme-catalysed reaction, this technique is extremely sensitive and can detect proteins in the pg (10^{-12} g) range. In a similar application, antibodies may be used to detect protein made from cloned DNA in so-called expression vectors (section 2.3). Further, a protein may be purified by immunoaffinity chromatography by using a specific antibody coupled to a suitable matrix (section 2.2.4e).

2.3 Recombinant DNA methods

The development since the 1970s of a wide range of new techniques now allows biochemists to study genes in detail at the level of the DNA, rather than at the level of the gene product (protein or RNA). This *recombinant DNA technology* is based on a few key principles:

(a)   DNA is cut at specific sites by enzymes called *restriction endonucleases*. These enzymes frequently recognise a specific palindromic sequence

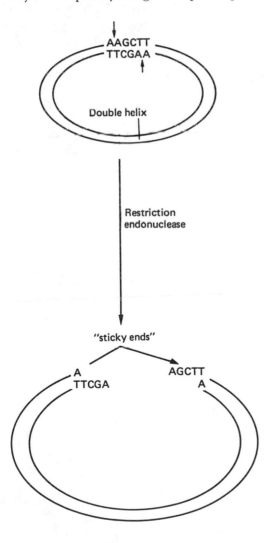

Figure 2.12
Action of restriction endonucleases

The action of the enzyme *Hind*III on a circular plasmid DNA is shown. The enzyme recognises a specific palindromic sequence (which reads AAGCTT in the $5' \rightarrow 3'$ direction on each strand), and cuts the two strands in a staggered fashion (at the sites indicated by arrows). The resulting linear DNA has cohesive termini ('sticky ends') with complementary base sequences.

Plasmid: Small (about 5 kb) circular self-replicating DNA found in bacteria and yeast. Often contains genes for resistance to certain antibiotics.

Intron: A non-coding intervening sequence of DNA that separates two DNA sequences in a gene which codes for protein or RNA.

Expression vector system: Genetically engineered plasmid or viral DNA with a site for inserting foreign DNA, in such a way that the cloned DNA can be transcribed with high efficiency. The bacterial host for the vector has low levels of proteases, so that proteins specified by the mRNA are not degraded extensively.

of bases, to give rise to fragments that have either blunt ends or staggered ('sticky') ends (figure 2.12).

(b) DNA fragments can be incorporated into bacterial **plasmids**, and culturing of the host bacterial cells will give rise to *clones*, each containing a single type of DNA insert (recombinant DNA).

(c) Single-stranded DNA will *hybridise* (by base-pairing) with a polynucleotide chain (either DNA or RNA) of complementary base sequence.

Genes may be cloned in the form of a *complementary DNA (cDNA) library* or a more complex *genomic (chromosomat) DNA library*. Since cDNA is prepared from messenger RNA, it lacks the **introns** and the flanking control regions that are present in genomic DNA.

The way in which these principles are exploited in order to generate a cDNA library is illustrated in figure 2.13. An individual clone has to be identified somehow, by using a suitable '*probe*'. If one knows part of the amino acid sequence of a given protein, then one can synthesise an *oligonucleotide probe* with the corresponding base sequence, as determined by the genetic code. Alternatively, certain **expression vectors** can be induced to synthesise (or express) the protein coded for by the inserted cloned DNA; the protein is detected with a specific antibody.

What does one do with a cloned cDNA? One can use it for *Southern blotting*, a technique named after its originator, **Ed Southern**. Here, genomic (chromosomal) DNA is digested with a restriction endonuclease, and the resulting fragments separated on the basis of their size by electrophoresis in agarose gel (section 2.2.3).

Figure 2.13
Preparation of cloned cDNA

Any cDNA with the appropriate 'sticky ends' can be integrated into a plasmid cut with the corresponding restriction endonuclease. The recombinant bacteria are selected on the basis of their resistance or sensitivity to certain antibiotics (since insertion of exogenous DNA inactivates one of two abtibiotic-resistance genes on the plasmid). The 'probe' for an individual clone could be an oligonucleotide (with a base sequence predicted from the amino acid sequence of the protein of interest), or a monospecific antibody to the protein (in expression vectors which synthesise recombinant protein). In practice, one would repeat the cloning step ('sub-clone') before analysing the DNA.

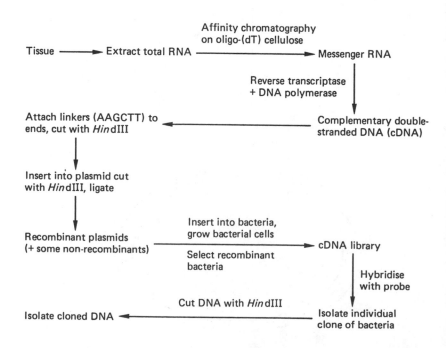

The fragments are transferred ('*blotted*') onto a nitrocellulose (or nylon) membrane, and probed with radioactively labelled cDNA by hybridisation. Fragments of DNA containing the gene concerned, and often its associated introns and regulatory sequences also, can thereby be identified (see, for example, figure 2.9). Alternatively, the mRNA corresponding to a cloned cDNA can be identified in an analogous procedure called *Northern blotting*. Furthermore, there are now reliable methods for the determination of the sequence of bases in a cloned fragment of DNA. Computers can be programmed to search such sequences for coding and non-coding regions, the frequency of each codon as a proportion of the total, and homology with other sequences. The coding sequence of the gene for a protein will give, *via* the genetic code, the corresponding amino acid sequence. In fact, it is now easier to sequence a protein through its DNA than through conventional techniques of protein chemistry. From the cloned cDNA, one can alter a single base in the sequence, and persuade an expression vector to synthesise large amounts of the modified protein for further study of its structure-function relationships — a technique called *site-directed mutagenesis*. Finally, one can determine the chromosomal location of a gene by hybridisation with a radioactively labelled cDNA probe (see, for example, figure 2.14).

Of what value is recombinant DNA technology to the study of subcellular organelles? We shall see numerous applications of these techniques in the chapters that follow, and even more in the next few years.

2.4 Summary

Information about the structure and function of subcellular organelles has been derived from three broad areas of experimental work. Imaging methods, such as electron microscopy and X-ray diffraction, visualise the morphology of cells, organelles, and the atomic architecture of

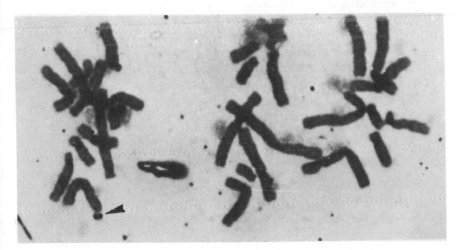

Figure 2.14
Locating genes on chromosomes

The locus of the gene for Factor VIII (a blood-clotting protein) is revealed by *in situ* hybridisation. The radioactive probe binds to the X-chromosome (arrowhead) at the position of the gene. [Taken from Buckle, V. J. and Craig, I. W. (1986), in *Human Genetic Disease – a Practical Approach* (Davies, K., ed.), IRL Press Ltd, Oxford; reproduced by permission of the publisher.]

macromolecules. Biochemical and immunochemical methods allow the researcher to fractionate the contents of cells and of each class of organelle, often in order to study a single component of a complex mixture. Centrifugation provides a means of preparing adequate amounts of purified organelles for subsequent analysis. Monoclonal antibodies have revolutionised immunochemistry; their ability to discriminate a single site on a single protein (or other macromolecule) underlies many highly specific and sensitive detection methods. Finally, recombinant DNA technology allows cell biologists to study individual components of organelles at the level of the genes concerned. We still lack information, however, on the three-dimensional structure of organelles and their dynamic interactions within the cell.

2.5   Study questions

1. An early model of membrane structure had a lipid bilayer coated on each face with protein in an extended β-pleated sheet conformation. What later experimental evidence has led to rejection of this model? (section 2.1.2)

2. A sample contains three proteins with the following properties: Protein A, pI = 7.0, mass = 68 kDa; Protein B, pI = 7.0, mass = 17 kDa; Protein C, pI = 5.0, mass = 17 kDa. How would you separate the three proteins from one another? (section 2.2.4)

3. When assayed for certain marker enzymes, peroxisomes do *not* show the property of structure-linked latency. What does this result suggest? (section 2.2.1)

4. Alkaline phosphatase is localised in the plasma membrane. After differential centrifugation of an isotonic liver homogenate, the enzyme is found in high concentrations in both the nuclear fraction and the microsomal fraction. Interpret these results. (section 2.2.1)

5. Growth hormone is a polypeptide synthesised by the pituitary gland. Outline the sequence of steps you would follow in order to prepare large amounts of human growth hormone by means of recombinant DNA technology. (section 2.3)

6. A polyclonal antiserum raised against pure adult haemoglobin cross-reacts with foetal haemoglobin, whereas a monoclonal antibody does not. Suggest a possible explanation. (section 2.2.7)

7. Glycoproteins containing α-D-mannose residues bind to a plant lectin (carbohydrate-recognition protein) called concanavalin. How could you use concanavalin for the purification of glyco-proteins by affinity chromatography? (section 2.2.4e)

2.6   Further reading

Additional information on techniques in cell biology is given in chapter 4 of Alberts *et al.* (1983). *Molecular Biology of the Cell*, Garland Publishing Inc., New York. A more comprehensive treatment is provided by the following:

Wilson, K. and Goulding, K. H. (1986). *Principles and Techniques of Practical Biochemistry*, 3rd edn, Edward Arnold, London.

Individual experimental aspects are covered in the following:

(a) Imaging

Crewe, A. V. (1971). *Sci. Amer.*, **224**, 26–35.
   (Scanning EM)
Bragg, L. (1968) *Sci. Amer.*, **219**, 58–70.
   (X-ray crystallography)

(b) Centrifugation

de Duve, C. (1975). *Science*, **189**, 186–194.
   (Exploring cells with a centrifuge)

(c) Electrophoresis and chromatography

Many of the techniques described for proteins are elaborated in

Jakoby, W. B. (1984). *Methods in Enzymology*, **104**, 1–503.

(d) Immunochemistry

Milstein, C. (1980). *Sci. Amer.*, **243**, 56–64
   (Monoclonal antibodies)

(e) Recombinant DNA

Steel, C. M. (1984). *Lancet*, (**ii**), 908–911 and 966–968.
   (The tools of recombinant DNA methods)
Anderson, W. F. and Diacumakos, E. G. (1981). *Sci. Amer.*, **245**(1), 60–93.
   (Genetic engineering and mammalian cells)

3 THE NUCLEUS

3.1 The nucleus: control centre of the cell

We saw earlier (chapter 1) how the nucleus was the first organelle to be studied. Its size (up to 10 μm roughly in diameter) and its characteristic staining properties meant that it was readily visible to the early light microscopists. Furthermore, it could be seen to be duplicated during the process of cell division. However, the true functional significance of the nucleus was only appreciated when it was realised that nuclear DNA is the genetic material. Since the DNA directs the biosynthesis of the cell's proteins and enzymes, and through them that of all the other macromolecules present in and around the cell, one could amply justify the assertion that the nucleus is the control centre of the cell. Despite this functional significance, we know relatively little about the structural organisation of the nucleus.

3.1.1 The nucleus contains most of the cell's DNA

Each human liver cell contains about 5 pg (5×10^{-12} g) of DNA, present in the form of *chromosomes* (figure 3.1). Each chromosome

Figure 3.1
Eukaryotic chromosomes

Micrographs of pairs of human chromosomes 1–22 and X are arranged into groups (A–G) on the basis of size. The characteristic banding pattern on each chromosome results from the Giemsa stain procedure. [Courtesy of E. Therman.]

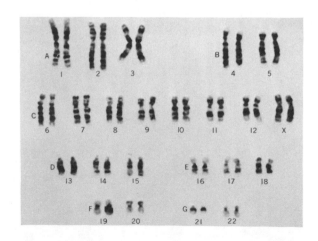

contains on average 10^8 base pairs, arranged in the familiar *double helix* first described by **Watson** and **Crick**. If the polynucleotide chain were present in its fully extended form, it would be roughly 3.4 cm long. Clearly it is not; we will return to this point (section 3.4). In fact, during **interphase** the DNA is relatively extended, and the chromosomes are not readily visible. In this state, the DNA is said to exist as *chromatin* (figure 3.2). The chromatin and the aqueous phase of the nucleus together constitute the *nucleoplasm*. The negatively charged polynucleotide chain of DNA exists in almost all eukaryotes in close association with positively charged proteins, the *histones* (section 3.4.1). Most double-stranded DNA is present under physiological conditions in the form of the *B-helix* structure (figure 3.3); in certain situations (for example, DNA–RNA hybrids), another right-handed double helix (the *A-helix*) may be favoured; rarely, an alternating sequence of purines and pyrimidines stabilises a much more open, left-handed structure, the *Z-helix* (figure 3.3).

The methods of recombinant DNA technology (section 2.3) have provided us with a great deal of information on the organisation of DNA. We do not have the space to present a detailed account here;

Interphase: The apparently quiescent period of the *cell cycle*, between cell divisions. It comprises the G_1 phase, prior to the onset of DNA synthesis during the S phase, and following this, the G_2 phase. Interphase precedes the M phase, which is characterised by nuclear division (mitosis) and cytoplasmic division (cytokinesis).

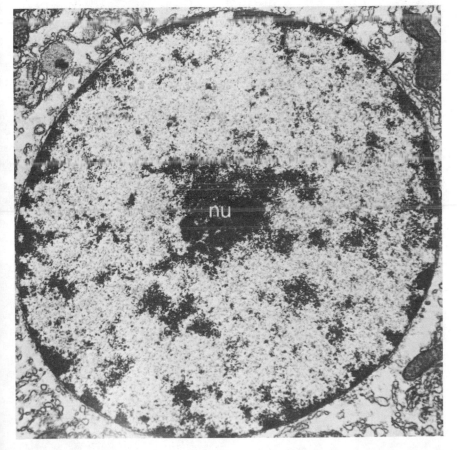

Figure 3.2
Morphology of the nucleus

This electron micrograph shows the nucleus of a liver parenchymal cell. Note the double membrane of the nuclear envelope, punctured in places by nuclear pores (arrows). The nucleoplasm is filled with extended chromosomes (chromatin); condensed chromatin, which appears electron-dense, is confined mainly to regions underlying the nuclear envelope. A single nucleolus (nu) is centrally placed. × 10 000. [Taken from Hopkins, C. R. (1978). *Structure and Function of Cells*, W. B. Saunders Co., Philadelphia; reproduced by permission of the publisher.]

Figure 3.3
Space-filling models of DNA

The B-helix (left) is right-handed, has major and minor grooves, and has about 10.4 base pairs per turn. The Z-helix (right) is left-handed, has a single type of groove that extends into the heart of the molecule, and has 12 base pairs per turn. (Heavy lines indicate the sugar–phosphate backbone; grey shading shows the bases accessible only in Z-DNA.) [Taken from Furth, A. and Moore, R. (1986). *Self Assembly of Macromolecules* (Book 2 of the Open University course S325, *Biochemistry and Cell Biology*), copyright 1986, The Open University Press.]

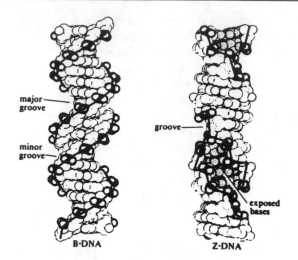

the reader should consult the reviews referred to in section 3.8, or consult another volume in this series (*What is a gene?*, by C. J. Skidmore). However, we will describe the recent major re-evaluation of the concept of the *gene,* partly in order to introduce certain terms that recur throughout this book. 'One gene — one polypeptide chain' is now an inadequate dogma; one gene may code for several functional proteins, and several spatially separated sequences of DNA may give rise a functional messenger RNA. The genetic code is not totally 'universal'; exceptions to the established codon usage occur in the mitochondria of animals and fungi (table 3.1). Very few eukaryotic genes are continuous nucleotide sequences: most are interrupted by intervening non-coding sequences, or *introns.* The regions corresponding to the introns are spliced out of the *precursor mRNA* (the primary transcript) by a mechanism of *RNA processing.* The mature messenger is covalently modified at each end by addition of a 7-methylguanosine '*cap*' linked via a triphosphate group to the 5'-terminal nucleotide, and up to 200 adenylate residues [*poly(A)* '*tail*'] to the 3'-terminus. Previously, it was thought that each gene was relatively fixed in position at its *locus* on the chromosome. It is now known that certain genetic elements, in particular the *transposons,* possess a structure that endows them with a high degree of mobility within the *genome.* Certain genes

Table 3.1
Exceptions to the 'universal' genetic code

Codon	'Universal' code	Yeast mitochondria	Mammalian mitochondria
UGA	Stop	Tryptophan[a]	Tryptophan[b]
AUA	Isoleucine	Methionine	Methionine[b]
CUA	Leucine	Threonine	Leucine
AGA	Arginine	Arginine	Stop

[a] Also true for the mitochondrial system of the fungi, *Neurospora* and *Aspergillus.*
[b] Also true for the mitochondrial system of the fruit fly, *Drosophila.*

may be classified as **structural** or as **regulatory.** Finally, the "central dogma of molecular biology" ('DNA makes RNA makes protein') is not totally valid: **reverse transcriptase** (RNA-directed DNA polymerase) can synthesise DNA from an RNA template.

Not all the cell's DNA is located in the nucleus. Mitochondria contain circular DNA that resembles the prokaryotic genetic material and represents about 1 per cent of the cell's total DNA (section 8.3.1). In photosynthetic plant cells, the chloroplasts also have a complement of DNA that codes for certain components of that organelle (section 9.5.3). We shall see later how this division of genes between nucleus and organelle creates certain problems in the biogenesis of mitochondria and chloroplasts.

3.1.2   The nucleus has a double membrane

Clearly visible on electron micrographs (as in figure 3.4) is the *double membrane* that delineates the nucleus. Between the inner and the outer membrane is a lumen (20–40 nm wide) referred to as the *perinuclear space*. The lipid composition of the nuclear membranes is unexceptional (table 1.5). We shall consider the structure and function of the nuclear membranes in more detail in section 3.2. We shall also see how the membranes undergo a process of reversible breakdown during cell division (section 3.5.3).

Between the inner membrane and the extremities of the chromosomes can be seen (in some cells) a dense fibrous protein meshwork, the *nuclear lamina*. Puncturing the nuclear membranes and lamina are nuclear pores, visible in figure 3.4 as structures where the inner and outer membranes make contact. The double membrane, lamina and pores

Structural gene: One which codes for a protein or tRNA or rRNA.
Regulatory gene: One which controls the expression of a structural gene.
Reverse transcriptase: An RNA-directed DNA polymerase present in certain viruses (including the AIDS virus).

**Figure 3.4
Structure of the nuclear envelope**

In this freeze-fracture electron micrograph, the plane of cleavage has followed the inner nuclear membrane and then has broken across the nuclear envelope and into the cytoplasm. The two membranes, the intervening perinuclear cisterna, and several pores (at arrows) are therefore seen in cross-section. In the upper part of the micrograph, a face view of the nuclear envelope shows pores randomly distributed over the surface. [Taken from Fawcett, D. W. (1981). *The Cell*, 2nd edn, W. B. Saunders Co., Philadelphia. Original micrograph courtesy of B. Gilula.]

nuclear pore

inner nuclear membrane

outer nuclear membrane

endoplasmic reticulum

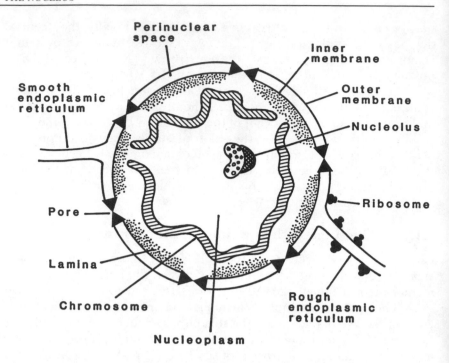

together constitute the *nuclear envelope*. A diagrammatic representation
of all the components of the nucleus can be seen in figure 3.5.

3.1.3   The presence of the nuclear membranes has important implications for gene expression

In eukaryotes, *transcription* of a functionally active gene generates a
mature mRNA in the nucleus. In order for the mRNA to act as a
template for *translation* (protein biosynthesis), it must gain access to
the ribosomes present in the cytosol. Thus, the nuclear membrane acts
as a physical barrier that separates these two processes in time and space.

A major consequence of this temporal and spatial separation is
that the pre-mRNA can undergo considerable modification before it
achieves its mature form, as mentioned in section 3.1.1. In some cases,
differential RNA processing may generate functionally distinct messengers
from the same primary transcript. For example, figure 3.6 shows how
the polypeptide hormone calcitonin and a protein called calcitonin
gene-related peptide (CGRP) are synthesised in the C-cells of the
thyroid gland and in the hypothalamus respectively. Here, differential
splicing of pre-mRNA gives rise to tissue-specific gene expression.

The retention of the primary transcript within the nucleus is clearly
essential for complete processing to generate the mature mRNA, but
it may also lead to the destruction of many potential messengers. When
cells were incubated with radioactive uridine, it was rapidly incorporated
into pre-mRNA in the nucleus. In the following 'chase' period with
non-radioactive medium, the subcellular location of the labelled
material was investigated. Over 95 per cent of the radioactivity

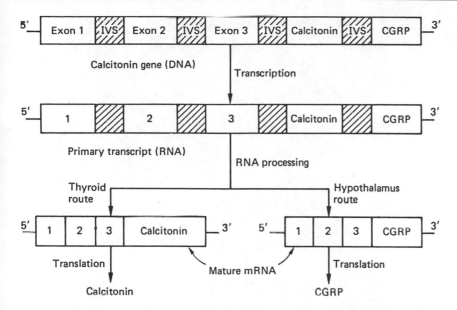

Figure 3.6
Differential processing of messenger RNA

The calcium-regulating hormone, calcitonin, and the calcitonin-gene related peptide (CGRP) are both derived from the same primary transcript. Tissue-specific RNA processing generates the mature mRNA corresponding to either the one protein or the other, by splicing of the appropriate 3'-exon. [IVS = intervening sequence (intron).]

remained in the nucleus, and was present ultimately in the form of oligonucleotides and degraded RNA. Less than 5 per cent of the primary transcripts emerged from the nucleus as mature mRNA. Little is known of the mechanisms whereby the nucleus allows only a proportion of all the messengers to leave and become templates for protein biosynthesis in the cytosol.

3.2 The nuclear envelope

We mentioned earlier (section 3.1.2) how the nucleus is surrounded by a double membrane, associated with pores and the underlying lamina. Here, we analyse the organisation of the nuclear envelope, and its role in the structure and function of this organelle.

3.2.1 The outer membrane is continuous with the endoplasmic reticulum

Electron micrographs have revealed the physical continuity that exists between the outer nuclear membrane and elements of the endoplasmic reticulum (as in figure 3.7). In principle, lateral diffusion should lead to intermingling of the lipid and protein components of the outer nuclear membrane and that of the endoplasmic reticulum. In practice, the nuclear envelope shares many enzymes with the endoplasmic reticulum, such as glucose 6-phosphatase, but their protein compositions are not identical. For example, they have different components of the cytochrome *P*450-dependent electron transport chain. Ribosomes are frequently observed in association with the outer nuclear membrane. Indeed, isolated nuclear envelopes contain all the components required for the biosynthesis of *N*-linked glycoproteins. It is on the basis of such

Figure 3.7
**Continuity between nuclear
envelope and endoplasmic
reticulum**

The outer nuclear membrane is
decorated with ribosomes. At one
point (large arrow) it is continuous
with the membrane of the rough ER.
× 62 000. [Taken from
Fawcett, D. W. (1966). *The Cell*,
W. B. Saunders Co., Philadelphia;
reproduced by permission of the
publisher. Original micrograph
courtesy of D. W. Fawcett.]

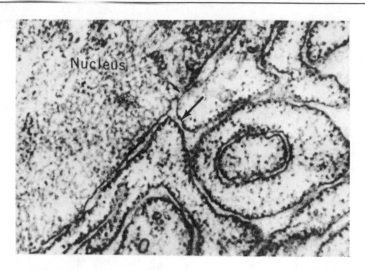

Lamina: Protein meshwork which
lines the nucleoplasmic face of the
inner membrane of nearly all
interphase nuclei.

evidence that the nuclear envelope and the rough endoplasmic
reticulum are considered to be functionally, as well as morphologically,
equivalent.

3.2.2   The inner membrane surrounds a fibrous protein meshwork, the lamina

In certain eukaryotic cells, electron microscopy of thin sections reveals
a discrete 30–100 nm thick layer between the inner nuclear membrane
and the chromatin. This layer is the **lamina,** which has been proposed
as an organising framework for the nuclear envelope, and as a site of
attachment for chromosomes.

If preparations of nuclear membranes are extracted with appropriate
non-ionic detergents and high-salt solutions, most of the membrane,
residual chromatin and nuclear pore complexes are removed. The
remaining protein, when analysed by SDS-PAGE, contains three major
polypeptides, with molecular masses of 60–75 kDa. These *lamins* (A, B,
and C) are immunologically related, and can be located by electron
microscopic immunocytochemistry as components of the nuclear lamina
(figure 3.8). The structure of two of the lamins (A and C) in
human somatic cells has been deduced from sequencing of cDNA clones.
They show strong sequence homology to one another, and to the
α-helical domain present in all intermediate-type filaments (IFs;
see section 1.4.2). Like the IFs, the lamins consist of a rod-shaped
coiled-coil dimer flanked by globular N-terminal and C-terminal
regions. Reconstitution of lamins A and C generates filaments with a
diameter of 10 nm, similar to the filaments of the intact nuclear lamina
of amphibian oocytes. In these cells, thin-section electron microscopy
reveals a meshwork of 10 nm filaments arranged at right angles to each
other (figure 3.9). The relatively thick (30–100 nm) nuclear lamina
of some cells presumably results from multiple layers of such meshworks.

The precise nature of the biological role of the lamina remains
uncertain. Although contacts between the lamina and chromosomes

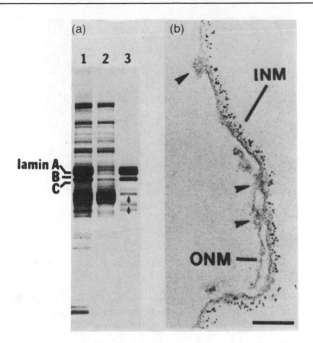

Figure 3.8
Organisation of the nuclear lamina

(a) Nuclear envelopes (lane 1) from rat liver were extracted with 2 per cent Triton X-100 and 0.3 M KCl, to give a soluble fraction (lane 2) and an insoluble fraction (lane 3). Protein components were analysed by SDS-PAGE. The extraction procedure separates the lamina network from nuclear membranes and pore complexes; the major contaminating proteins (arrows in lane 3) are intermediate filament polypeptides. (b) The lamins were located in isolated nuclear envelopes by immuno-electron microscopy using ferritin-labelled antibody. Labelling occurs only at the nucleoplasmic face of the inner nuclear membrane (INM). The fragile outer nuclear membrane (ONM) is detached in places. Arrows mark pore complexes. (Bar = 0.15 μm). [Taken from Gerace, L. (1986). *Trends Biochem. Sci.*, **11**, 443–446; reproduced by permission of Elsevier Science Publishers.]

can be visualised by electron microscopy, the functional significance of this association is unclear. Lamin B remains associated with membrane vesicles after the nuclear envelope disassembles during mitosis (described further in section 3.5.3). Perhaps the lamina, *via* its own reversible disassembly, orchestrates the disintegration and reformation of the nuclear membranes during cell division.

3.2.3 Pores in the nuclear envelope connect the nucleoplasm with the cytosol

Contact between the nucleoplasm and the cytosol is maintained *via* *nuclear pore complexes*, 60–100 nm in diameter (figures 3.4 and 3.10). There may be a few thousand pores in the nuclear envelope of a somatic mammalian cell, or a few million in the much larger amphibian oocyte.

In electron micrographs (figure 3.10a), the nuclear pore complex appears as an octagonal array of proteins with a total mass of about 10^5 kDa. Monoclonal antibodies have defined eight constituent polypeptides, but there are undoubtedly more. A postulated arrangement of the components of the complex is presented in figure 3.10b. The diameter of its aqueous channel may be as small as 10 nm or as large as 20 nm; even a cavity 20 nm across is too small to permit the passage of a 60S ribosomal subunit (diameter 25 nm). Presumably, there is a change in shape of either the pore or the ribosomal subunit in transit.

The nuclear pore facilitates a two-way traffic of material between nucleus and cytosol. The major components leaving the organelle include ribosomal subunits, transfer RNA and messenger RNA; those entering from the cytosol include nuclear proteins and enzymes, coenzymes, ATP and other relatively small solutes. At the same time,

Figure 3.9
Native nuclear lamina of amphibian oocyte

In this electron micrograph of a freeze-dried, metal-shadowed nuclear envelope extracted with Triton X-100, one can clearly see the nuclear lamina meshwork partially covered with arrays of nuclear pore complexes. The filaments that comprise the lamina appear to run at right angles to each other. (× 10 000). [Courtesy of Dr U. Aebi.]

the nucleus must be able to retain its constituent soluble proteins. What regulates the passage of a given component through the nuclear pore in one direction only?

Nuclear proteins contain a 'signal sequence' for entry into the nucleus from the cytosol. This conclusion was reached from studies with *nucleoplasmin,* a major soluble protein of the nuclei of *Xenopus* (amphibian) oocytes. When radioactive nucleoplasmin is microinjected into the cytosol, the protein accumulates in the nucleus. No transient larger precursor is involved in the uptake mechanism, in contrast to the processes that lead to incorporation of newly synthesised proteins of lysosomes, mitochondria and chloroplasts (sections 6.4.2, 8.4.1 and 9.5.4, respectively). Removal of the *C*-terminal domain of nucleoplasmin by partial proteolysis abolishes uptake; this fragment also accumulates in the nucleus after microinjection. In related experiments, a cytosolic protein (the glycolytic enzyme, pyruvate kinase) could be selectively taken up by nuclei if it was first fused (by recombinant DNA experiments) to the large T antigen protein of the virus SV40. The minimal sequence of the viral antigen that facilitated nuclear uptake of the fusion protein was Pro.Lys.Lys.Lys.Arg.Lys.Val. This sequence shares homology with that of the histones and two yeast nuclear proteins (MAT α1 and α2), but it is not yet known whether all nuclear proteins share a common signal sequence. Likewise, the mechanism whereby such proteins pass selectively through the nuclear pore complex remains uncertain. Colloidal gold particles (5–20 nm diameter) coated with nucleoplasmin can be visualised by electron microscopy during their passage through the pore's orifice (figure 3.11). However, the structural features that permit the selective exit of products generated by the nucleus, presumably through the pores, have not yet been defined.

3.3   The nucleolus: site of the biogenesis of ribosomes

The nucleolus was visible to the early microscopists as a darkly staining patch (or patches) within the nucleoplasm. We now know that the only function of this 'organelle' is the generation of ribosomes. Its

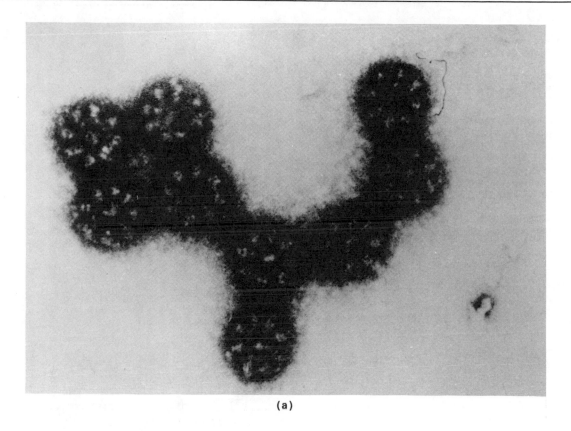

(a)

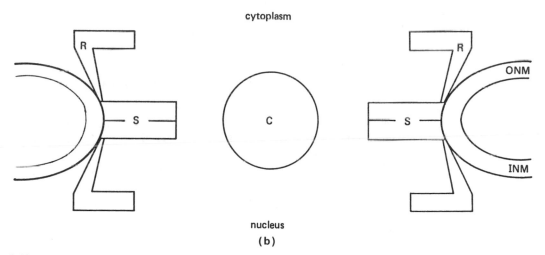

cytoplasm

nucleus

(b)

Figure 3.10
Structure of the nuclear pore complex

(a) Pore complexes were released onto the electron microscopy grid from a nuclear envelope immersed in low-salt medium containing 0.1 per cent Triton X-100. Note the ring-like structures composed of eight particulate components. A central 'plug' is invariably present. × 75 000. (b) Model of a pore complex in cross-section. The major components are the rings (R), spokes (S), and central plug (C). The outer and inner nuclear membranes (ONM and INM respectively) come together at the external margin of the spokes; the distance between these points is about 60 nm. [Taken from Unwin, P. N. T. and Milligan, R. A. (1982). *J. Cell Biol.*, **93**, 63–75; reproduced by copyright permission of the Rockefeller University Press.]

Figure 3.11

Transport into the nucleus *via* the pore complex

Colloidal gold particles coated with the protein nucleoplasmin were micro-injected into the cytoplasm of an amphibian oocyte. In this electron micrograph, the particles can be seen migrating from the cytoplasm (C) through a pore complex into the nucleus (N). In a control experiment, particles coated with a non-nuclear protein remained in the cytoplasm. [Taken from Feldherr, C. M., Kallenbach, E. and Schultz, N. (1984). *J. Cell Biol.*, **99**, 2216–2222; reproduced by copyright permission of the Rockefeller University Press.]

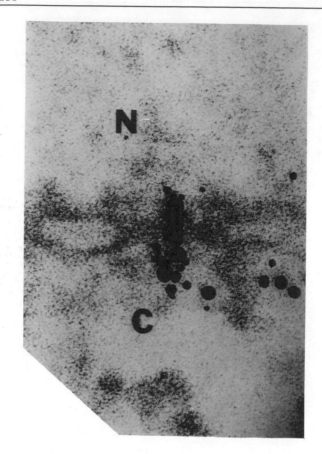

structure develops and disassembles during the course of the cell cycle, such that its activity is finely tuned to the metabolic demands of the cell.

3.3.1   The nucleolus forms a morphologically distinct region of the nucleoplasm

Electron micrographs (such as figure 3.12) of the nucleolus show a heterogeneous structure. There is no limiting membrane. Rather, specific nucleolar components generate regions of electron opacity that appear fibrillar or granular. A diagrammatic representation of the ultrastructural organisation of the nucleolus can be seen in figure 3.13. The likely structure and function of its components are indicated in table 3.2. The initiating sites of nucleolar organisation are the genes that encode ribosomal RNA. Tandemly arranged blocks of rRNA genes, located on various chromosomes, associate with histones and non-histone proteins to form ribosomal chromatin, which in turn acts as the *nucleolar organising regions*. The ribosomal chromatin forms the fibrillar centres, whose transcriptional activity generates the ribosomal precursors that successively form the dense fibrillar component and then the granular component. The nucleolus may interact structurally and functionally with the surrounding chromatin, or with the nuclear envelope (*via* lamellar or tubular lipoprotein structures).

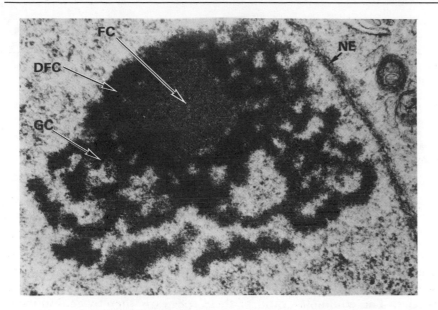

Figure 3.12
Morphology of the nucleolus

The major features of the nucleolus are visible in this electron micrograph of a thin section through a mammalian nucleus: FC, fibrillar centre; DFC, dense fibrillar component; GC, granular component. Their structure and function are outlined in table 3.2. NE, nuclear envelope. [Taken from Fawcett, D. W. (1981). *The Cell*, 2nd edn, W. B. Saunders Co., Philadelphia; reproduced by permission of the publisher. Original micrograph courtesy of D. W. Fawcett.]

3.3.2 Ribosomal subunits are assembled in the nucleolus

The 60S (large) and 40S (small) subunits of eukaryotic ribosomes together contain one copy each of 28S, 18S, 5.8S and 5S rRNA, and of about 80 distinct proteins. How does the nucleolus synchronise the formation and assembly of equimolar amounts of the rRNA species and of the ribosomal proteins?

Activation of ribosomal chromatin leads to its progressive unfolding so as to expose transcriptionally active units. Transcription catalysed

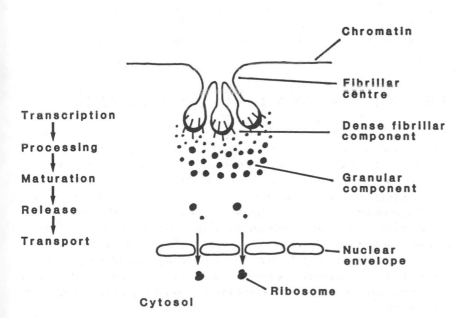

Figure 3.13
Structure and function of the nucleolus

In this schematic view, the morphological features visible under the electron microscope are indicated on the right. On the left are the processes involved in the biogenesis of ribosomes. Processing of the pre-rRNA and association of rRNA with ribosomal proteins (maturation) generates 40S and 60S ribosomal subunits.

	Component	Structure	Function
Table 3.2 **Components of the nucleolus**	Fibrillar centre	5 nm fibrils plus condensed DNA	Association of nucleolar organising regions (genes for rRNA)
	Dense fibrillar component	4–10 nm fibrils, electron-dense	Transcription of rRNA genes, assembly of pre-ribosomal particles
	Granular component	15–20 nm particles	Processing and maturation of pre-ribosomes

by RNA polymerase I generates pre-ribosomal particles containing a large primary transcript, the 45S RNA. This precursor species is selectively cleaved by several specific endonucleases acting in a strict sequence so as to form one copy each of the 28S, 18S and 5.8S rRNA species. The equimolar ratio of these species is thereby maintained. This explanation does not take into account the 5S rRNA, which is synthesised by a different enzyme, RNA polymerase III. The genes for the 5S rRNA and the ribosomal proteins, like that of the 18S–5.8S–28S rRNA, seem to respond to common growth-dependent factors regulating their transcription.

Ribosomal proteins associate with nascent rRNA transcripts and with pre-ribosomal particles in the nucleolus and in the nucleoplasm. The precise pathway for this undoubtedly complex process has not yet been determined. Immuno-electron microscopy has identified the largest polypeptide (S1) of the 40S subunit as having a location in the granular component. This protein is therefore added at a relatively late stage of ribosomal biogenesis. Similar studies with other ribosomal proteins (such as S14) indicate a location in the fibrillar centre, and presumably an association with a nascent rRNA at an early stage of the process.

3.4   How is DNA packaged within the nucleus?

We referred earlier (section 3.1.1) to the 'packaging problem': how a total of 1.5 m of mammalian DNA could be accommodated within a nucleus less than 10 μm in diameter. In terms of volume, the DNA of the interphase chromosomes has been condensed by a factor of 1000 or so. The inherent rigidity of the double helix, and its negative charge, place considerable constraints on the ability of DNA to fold into more compact structures. Nevertheless, such folding is achieved. Yet for proteins, such as RNA polymerase or regulatory factors, to gain access to specific base sequences, the highly condensed DNA has to be unfolded at specific sites. Clearly, DNA folding is intimately linked with gene expression.

3.4.1 Folding of DNA generates successive higher-order structures

By incubation of isolated nuclei at low ionic strength, and with suitable preparation techniques, unwound chromosomal DNA can be visualised under the electron microscope (figure 3.14). One striking feature is the '*beads-on-a-string*' structure. The 'beads' represent *nucleosomes*, which are connected by *linker DNA* (the 'string'). Cleavage of the linker DNA by extensive digestion with micrococcal nuclease permits the isolation of intact nucleosomes, by density-gradient centrifugation. The nucleosomal DNA forms about two turns of a left-handed superhelix, wound around an octamer of histone proteins. X-ray diffraction studies suggest a disc-shaped particle, 5.7 nm thick and 11 nm in diameter. The double helix is not bent uniformly, but has regions of tight bending that are in contact with histones H3 and H4. A diagrammatic representation of the nucleosomal structure is shown in figure 3.15. This level of DNA folding corresponds to a roughly seven-fold condensation.

The 'beads-on-a-string' structure of chromatin is thought to be wound into a solenoid configuration, with six nucleosomes per turn, so as to generate the '*30 nm fibre*' that can be identified under the electron microscope. Histone H1 is essential for the stabilisation of this particular chromatin structure. This degree of condensation still represents only an approximately 40-fold compaction of the DNA. Further folding of the chromatin generates *looped domains* like those seen in lampbrush chromosomes (section 3.4.2), but there must exist yet higher-order structures before we arrive at the highly condensed

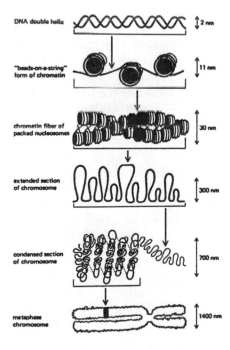

DNA double helix — 2 nm

"beads-on-a-string" form of chromatin — 11 nm

chromatin fiber of packed nucleosomes — 30 nm

extended section of chromosome — 300 nm

condensed section of chromosome — 700 nm

metaphase chromosome — 1400 nm

Figure 3.14
Levels of DNA packing

This schematic diagram illustrates the increasing orders of DNA packing that give rise ultimately to the highly condensed metaphase chromosome. [Taken from Alberts, B. *et al.* (1983). *Molecular Biology of the Cell*; reproduced by permission of Garland Publishing Inc.]

Figure 3.15
Structure of the nucleosome

The 1.8 turns of DNA go through a 'kink' at their mid-point, and extend into the linker DNA (not shown). Individual subunits of the histone octamer, $(2A, 2B, 3, 4)_2$, are indicated.

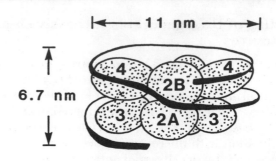

metaphase chromosome. The structural relationships between these successive levels of DNA folding are illustrated in figure 3.14.

3.4.2   Gene expression involves unfolding of the condensed DNA

Several lines of experimental evidence indicate that DNA is not expressed if it is present in a highly condensed form. Conversely, gene expression is only possible if the folded double helix is unravelled.

(1)   Somatic mammalian cells of female individuals have only one active X chromosome. Its inactive homologue is present as a highly condensed **heterochromatic** structure, the *Barr body*.

(2)   *Position-effect variegation*: if a transcriptionally active gene in an **euchromatic** region is translocated (by a mutational event) to a heterochromatic region, then gene expression is terminated. The reverse process also occurs.

(3)   The salivary gland cells of the *Chironomus* fly have enormously amplified chromosomes, in which the DNA is duplicated many times along the length of the chromosome. Differential DNA coiling results in characteristic banding patterns. Nuclei labelled with [³H]-uridine and subjected to autoradiography show transcriptional activity associated with regions of unwound DNA, the *chromosome puffs*.

(4)   *Lampbrush chromosomes* appear in amphibian oocytes during meiosis. The lateral loops along their length are hightly active in mRNA synthesis. Digestion of lampbrush chromosomes with proteases removes all associated protein from the chromatin. Such treatment leaves a polynucleotide strand in each loop that has the thickness (2 nm) expected for a bare, unfolded double helix.

Heterochromatin: Region of a chromosome that stains darkly with the dyes used for light microscopy, or the stains used for electron microscopy (as in figure 3.2).

Euchromatin: Lightly staining region of a chromosome.

Thus, there is convincing evidence that the degree of folding or unfolding of the chromatin may influence, at a gross level, the expression of certain genes. There are obviously more subtle levels of control that regulate the transcription of specific eukaryotic genes, but such mechanisms fall outside the scope of this book.

3.5   The nucleus and cell division

By definition, almost all of the foregoing account pertains to the nucleus during *interphase*. However, it is during *mitosis* that the nucleus undergoes a spectacular change in its structure as the cell undergoes division. The events involved are readily visible under the light microscope (as shown in figure 3.16). A major feature of this process is the reversible disassembly of the nuclear apparatus.

3.5.1   Chromosomes undergo duplication and further condensation during cell division

All the considerations relating to unfolding of the DNA during transcription (section 3.4.2) apply equally to DNA replication. DNA polymerase catalyses the formation of daughter DNA at each of the multiple replication forks along the chromosome. Histones labelled radioactively prior to replication remain associated with only one of the daughter double helices; the other has to acquire the appropriate histones before it can reform nucleosomes and re-fold. The net result of replication is the lateral association of two highly condensed sister **chromatids** joined at their **centromeres**. This metaphase stage of mitosis can be halted by incubation of cells with colchicine (section 1.4.2). The resulting display of the stained chromosomes, with their characteristic shapes and sizes, is referred to as the *karyotype* (figure 3.1). Karyotyping may reveal the underlying defect behind certain genetic disorders. For example, most individuals with Down's syndrome have an additional copy of chromosome 21 (trisomy 21); and Burkitt's lymphoma, a cancer affecting B cells (lymphocytes), is

Chromatid: One of the pair of daughter chromosomes generated by replication.
Centromere: Pronounced constriction in the chromosomal structure.

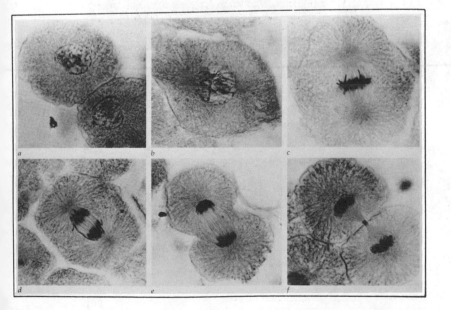

**Figure 3.16
Partition of chromosomes during cell division**

These light micrographs show the process of mitosis in cells from a whitefish embryo: (a) interphase; (b) prophase; (c) metaphase; (d) early anaphase; (e) late anaphase; (f) telophase. Note how the nuclear envelope begins to disappear during prophase, and how a fibrous array, the mitotic spindle, appears during metaphase. Only in late telophase does the spindle disappear, and the nucleolus and nuclear envelope re-form. [Taken from Macleod, A. G. (1975). *Cytology*, The Upjohn Company, Kalamazoo; reproduced by permission of the publisher.]

Figure 3.17
The mitotic spindle

In this electron micrograph, the
microtubules which form the spindle
originate from the region of the
centrioles (C). At this stage of mitosis
(metaphase), the chromosomes
orientate themselves along the equator
of the spindle. One chromosome has
been sectioned longitudinally in the
plane of its centromere (arrows).
× 10 000. [Taken from
Hopkins, C. R. (1978). *Structure and
Function of Cells*, W. B. Saunders Co.,
Philadelphia. Original micrograph
courtesy of B. R. Brinkley and
J. Cartwright.]

invariably associated with a translocation of part of chromosome 8 on
to the tip of chromosome 14.

3.5.2   The sister chromatids are partitioned by the mitotic spindle

Partition of the replicated DNA to each of the daughter cells involves
the alignment of the sister chromatids on a structure called the *mitotic
spindle* (figure 3.17). Evidence that microtubules (section 1.4.2) are an
essential component of this structure came from inhibition studies with
colchicine. This alkaloid drug blocks cell division because it interferes
with the association of microtubules that make up the mitotic spindle.
Under normal circumstances, this subcellular apparatus makes contact
at each pole with two cylindrical centrioles (section 1.4.2), and with

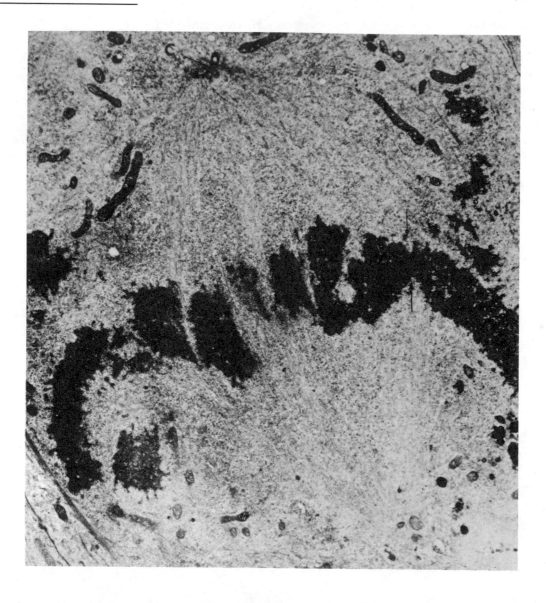

the chromosomal **kinetochores.** At the anaphase stage of mitosis, the sister chromatids separate and the individual homologous chromosomes migrate to opposite poles of the cell (figure 3.16). There is currently no model that accounts satisfactorily for all aspects of this movement of chromosomes.

Kinetochores: Granular structures associated with the centromeres of the chromatids.

3.5.3 The nuclear envelope and the nucleolus disassemble and reassemble during mitosis

During the prophase stage of mitosis, the nuclear envelope of most cells is seen to disappear when dividing cells are viewed under the light microscope; the nucleolus does likewise. Removal of the ribosomal chromatin for replication purposes presumably destroys the nucleolar organising regions, thereby leading to dissociation of the components that normally comprise the nucleolus (section 3.3.1). Ribosomal proteins, released as this structure disintegrates, remain associated with the chromosomes and integrate themselves into the post-mitotic nucleolus. Other nucleolar proteins can be seen by immunofluorescence microscopy to become dispersed throughout the cytosol.

The nuclear lamina and pores also undergo disintegration during mitosis. Lamins A and C become phosphorylated and are distributed to the cytosol, whereas lamin B appears to remain selectively associated with fragments of the nuclear envelope. It is postulated that lamina disassembly during prophase acts as a signal for the vesiculation of the nuclear membranes. Conversely, reassembly of the lamina during telophase probably initiates the fusion of these membrane vesicles to form a new nuclear envelope around the chromosomes. Evidence supporting this view has come from studies *in vitro* with a nuclear assembly system; depletion of lamin B inhibited the assembly of membrane and pore complexes around chromosomes.

During mitosis, disintegration of the nuclear envelope is accompanied by the formation of many small membrane vesicles (about 0.2 μm in diameter), visible under the electron microscope. Partitioning of these vesicles to the poles of the diving cell ensures a supply of precursors for reassembly of the new nuclear envelopes.

3.6 Summary

The nucleus, a large (5–10 μm diameter) organelle bounded by a double membrane, is the repository of the cell's genetic material, the chromosomal DNA. The outer membrane is continuous with the endoplasmic reticulum, whereas the inner membrane is overlaid on its inner surface with a fibrous protein meshwork, the lamina. The double membrane is punctured by nuclear pores, each constituted by a complex of proteins arranged in a ring-like structure. Nuclear proteins, imported from the cytosol in their mature form, probably have a signal sequence essential for entry *via* the pores. The nucleolus comprises a cluster of transcriptionally active rRNA genes that is associated with a reservoir of precursor ribosomal particles at various stages of maturation and processing. The remainder of the nucleoplasm contains double helical DNA packaged with histones into chromosomes. Condensation of the DNA involves its organisation successively into nucleosomes, 30 nm fibres, looped domains, and yet higher-order structures. During cell division, the components of the nucleus undergo extensive structural change, including disassembly/reassembly of the nuclear membranes and lamina, and of the nucleolus.

3.7   Study questions

1. From table 1.1, the nuclear membrane comprises about 0.2 per cent of the total cell membrane (about 10^5 μm^2 in a mammalian cell). If the nuclear membranes fragment during mitosis to form vesicles 0.2 μm in diameter, how many such vesicles would be formed? (Surface area $= 4\pi r^2$)

2. What evidence is there that unfolding of the condensed chromosomal DNA is an essential preliminary step in gene expression? (section 3.4.2)

3. In the histone octamer of the nucleosome (figure 3.15), how might one determine experimentally the relative positions in space of the individual histone monomers?

4. In mammalian and avian cells, lamin A is synthesised as a precursor about 2 kDa larger than the mature protein. How could one determine experimentally if the 2 kDa extension represented a signal sequence for targeting the protein to the nucleus?

5. Although it is assumed that mature mRNA passes out of the nucleus *via* the nuclear pores, rarely has this process been observed under the electron microscope. Devise an experiment that might lead to such an observation. (*Hint*: consider how the *import* of the nuclear protein, nucleoplasmin, has been demonstrated; see figure 3.11.)

6. From our knowledge of the composition of the nucleolus (section 3.3.1), suggest how one might determine experimentally which chromosomal regions contribute to its structure.

7. The nuclear envelope undergoes a continual increase in surface area during interphase in growing cells. Speculate on the mechanisms involved in expanding the nuclear membranes and the nuclear lamina.

3.8   Further reading

Chromatin structure and gene expression

Weisbrod, S. (1982). *Nature*, **297**, 289–295.
    (Active chromatin)

Igo-Kemenes, T., Hoerz, W. and Zachau, H. G. (1982). *Ann. Rev. Biochem.*, **51**, 89–122.
    (Active chromatin)

Chambon, P. (1981). *Sci. Amer.*, **244**(5), 60–71.
    (Split genes in eukaryotic DNA)

Darnell, J. E. Jr (1983). *Sci. Amer.*, **249**(2), 90–100.
  (RNA processing)
Brown, D. D. (1981). *Science*, **211**, 667–674.
  (Gene expression in eukaryotes)
Kornberg, R. D. and Klug, A. (1981). *Sci. Amer.*, **244**(2), 52–64.
  (Nucleosome structure)

Structure and function of non-chromatin components

Dingwall, C. (1985). *Trends Biochem. Sci.*, **10**, 64–66.
  (Import of nuclear proteins)
Gerace, L. (1986). *Trends Biochem. Sci.*, **11**, 443–446.
  (Nuclear lamina)
Sommerville, J. (1986). *Trends Biochem. Sci.*, **11**, 438–442.
  (Nucleolus and ribosomal biogenesis)

4 THE ENDOPLASMIC RETICULUM

4.1 Introduction

The endoplasmic reticulum (ER) may form up to half of the cell's total membrane (table 1.1). This proportion is strikingly high in those cells specialised for the export of lipids (such as hepatocytes and cells of the gonads) or of proteins (such as pancreatic cells or certain classes of B lymphocytes). However, almost all cells contain an extensive array of membranes that can be seen under the electron microscope to ramify throughout the cytosol (as in figure 1.3). Biochemical characterisation of the ER has been aided by the tendency of its membranes to fragment into small vesicles during during homogenisation of tissue samples. The resulting '*microsomes*' can be readily prepared by centrifugation (section 2.2.1). As we shall see, microsomes have been widely used as cell-free model systems for studying many of the biosynthetic processes mediated by the ER.

On the basis of their morphology, two principal classes of ER can be distinguished:

(1) The *rough endoplasmic reticulum*. This consists of nearly parallel arrays of cisternae, with ribosomes bound by their large (60S) subunit to the cytosolic face of the membranes (figure 4.1).

(2) The *smooth endoplasmic reticulum*. This consists mainly of an interconncted network of tubules, 80–120 nm in diameter (figure 4.2). It lacks ribosomes.

These structural distinctions are paralleled by functional differences. The rough ER is the site of synthesis of **secretory proteins,** whereas the smooth ER is concerned mainly with lipid metabolism and detoxification. Microsomes derived from rough ER and smooth ER can be separated by density-gradient centrifugation. The resulting 'rough' and 'smooth' microsomes have almost identical compositions, despite the striking functional differences between the parent membranes. There are, in addition, elements of the ER that appear to be intermediate, in terms of their morphology and degree of ribosomal decoration, between rough ER and smooth ER

Secretory proteins: Proteins secreted from the cell that must cross at least one membrane in transit. (Topologically, the extracellular space is equivalent to the lumen of the ER, of the Golgi complex, and of lysosomes. Hence, the term 'secretory proteins' will also be applied here to those proteins localised in these organelles.)

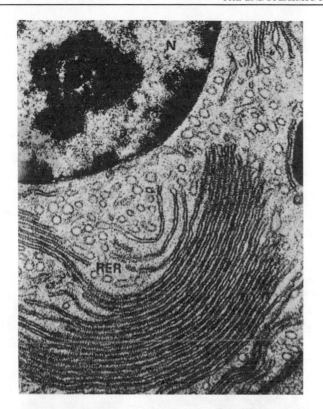

Figure 4.1
Morphology of the rough endoplasmic reticulum

Parallel arrays of cisternae, whose membranes are studded with ribosomes, are visible in the lower portion of this electron micrograph. Closer to the nucleus are circular and elliptical profiles where the cisternae have been cut through transversely or obliquely. RER, rough endoplasmic reticulum; N, nucleus. (Thin section of a pancreatic acinar cell.) [Taken from Fawcett, D. W. (1981). *The Cell*, 2nd edn, W. B. Saunders Co., Philadelphia; reproduced by permission of the publisher. Original micrograph courtesy of D. W. Fawcett.]

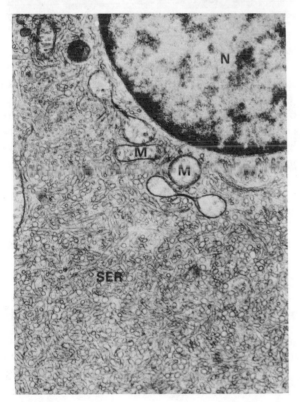

Figure 4.2
Morphology of the smooth endoplasmic reticulum

The smooth ER (SER) is particularly prominent in steroid-secreting endocrine gland cells. As seen in this electron micrograph, it consists of a network of tubules (80–120 nm diameter) extending through most of the cytoplasm. Experimental stimulation of steroid secretion causes a marked proliferation of the smooth ER. N, nucleus; M, mitochondria. (Thin section of a testicular Leydig cell.) [Taken from Christensen, A. K. and Fawcett, D. W. (1961). *J. Biophys. Biochem. Cytol.*, **9**, 653–670; reproduced by copyright permission of the Rockefeller University Press.]

Figure 4.3
Continuity of rough ER and smooth ER

In this electron micrograph, the cisternae of the rough ER (RER) are continuous with branched tubular elements (see arrows) which resemble smooth ER (SER). Numerous mitochondria (M) are also evident. (Thin section of mammalian liver cell.) [Taken from Fawcett, D. W. (1981). *The Cell*, 2nd edn, W. B. Saunders Co., Philadelphia. Original micrograph courtesy of R. Bolender.]

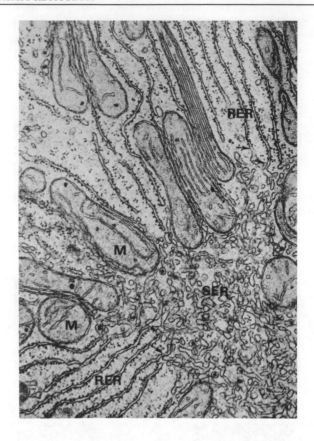

(figure 4.3). This *transitional endoplasmic reticulum* is continuous with the rough ER and the smooth ER, such that the whole membranous network of the endoplasmic reticulum encloses a single uninterrupted lumen, separated from the cytosol throughout by one membrane. A diagrammatic representation of this structure is shown in figure 4.4. By a process of budding and fusion, vesicles derived from the ER make contact with many of the cell's membrane-bound organelles. Also, the ER appears to be continuous with the outer nuclear membrane (section 3.2.1). Hence, processes started in the ER may continue at other intracellular sites, and so we cannot consider the ER in isolation from the other organelles of the cell.

The major function of the ER that is common to all cells is membrane biogenesis. Such a role is ideally suited to an organelle with a high capacity for synthesising both lipids and proteins, as we shall see later (section 4.4).

4.2   The rough endoplasmic reticulum

Early evidence for the biosynthetic role of the rough ER came from autoradiographic studies (section 2.2.2) of exocrine cells of the mammalian pancreas. Radioactive label, corresponding to newly synthesised secretory proteins, appeared first in the rough ER before

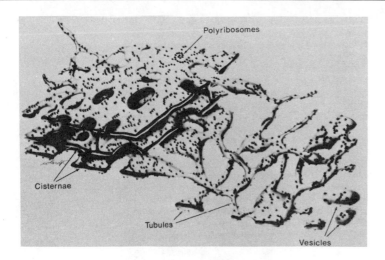

Figure 4.4
Three-dimensional organisation of the rough endoplasmic reticulum

Drawing of the interconnections between elements of the rough ER: fenestrated cisternae (left), branching tubules (centre), and vesicles (right). The cytoplasmic faces of the membranes are decorated with ribosomes. [Taken from Fawcett, D. W. (1981). *The Cell*, 2nd edn, W. B. Saunders Co., Philadelphia; reproduced by permission of the publisher.]

passing through the Golgi complex into secretory granules (figure 4.5). We now know in some detail the mechanism whereby cytosolic amino acids become incorporated into a polypeptide that can cross the membrane of the ER. Moreover, the **post-translational processing** (or modification) undergone by many secretory proteins is initiated in the rough ER.

Post-translational processing: Covalent changes in the structure of a newly synthesised polypeptide.

4.2.1 Most nascent secretory proteins have an N-terminal signal peptide

When mRNA from lymphocytes was subjected to translation *in vitro*, immunoglobulin light chain was synthesised as a precursor 4 kDa larger than the mature protein *in vivo* (figure 4.6, lane B). Addition to the experimental system of microsomes resulted in the conversion of the precursor polypeptide to its mature (smaller) form (figure 4.6, lane S). Furthermore, the mature protein was resistant to digestion by added protease, unless detergent was also added to the system (figure 4.6, lanes C and D). This intriguing experimental observation suggested that the processed immunoglobulin light chain was somehow shielded from the aqueous medium by a membrane.

Figure 4.5
Protein secretion by pancreatic exocrine cells

In pulse-chase studies with cells incubated with [^{14}C]-leucine, the migration of radioactively labelled proteins was followed by autoradiography and electron microscopy. This schematic diagram illustrates the experimental results. Label (indicated here by the symbol ⟋) first appears over the rough endoplasmic reticulum (RER). It later moves to the Golgi complex (G) and thence to secretory granules (S), before appearing in the extracellular fluid (not shown). The time elapsed, after the initial brief pulse with radioactive amino acid, is indicated. N, nucleus.

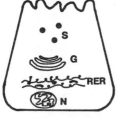

5 minutes   **7 minutes**   **10 minutes**

Figure 4.6
Synthesis of a secretory
protein

Lymphocyte mRNA was translated
in vitro and the resulting radioactive
polypeptides were resolved by SDS-
PAGE and autoradiography (as in
figure 2.10). Immunoglobulin (Ig)
light chain of 25 kDa (downward-
pointing arrow) was synthesised in the
presence of mRNA (lane B) but not
in the absence of mRNA (lane A).
Incorporation of microsomes into the
translation system (lane S) caused the
appearance of mature Ig light chain
of 21 kDa (upward-pointing arrows).
Subsequent addition of protease
(lane C) did not lead to degradation
of the Ig light chain, unless detergent
was also added (lane D). [Taken
from Blobel, G. and Dobberstein, B.
(1975). *J. Cell Biol.*, **67**, 852–862;
reproduced by copyright permission of
the Rockefeller University Press.]

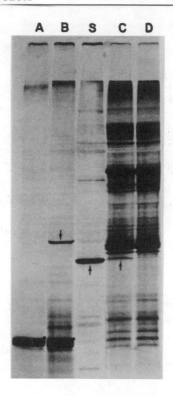

The interpretation of these results by **Blobel**, in the form of the
'*signal peptide hypothesis*', is now widely accepted. Most secretory proteins
have an *N*-terminal extension that directs them from a membrane-
bound ribosome to the lumen of the rough ER. This 'signal peptide'
(or sequence) is typically 15 to 30 amino acids long. Its composition
has been determined for many proteins, either directly from polypeptide
sequencing, or indirectly by deduction from the sequence of cloned
cDNA. Although they show no obvious sequence homology, the signal
peptides studied to date share a common feature: a high proportion
of hydrophobic amino acids (table 4.1). Such a structure would
facilitate an interaction with the lipid bilayer of the rough ER,
prior to the vectorial discharge of the nascent polypeptide across
the membrane. This process is termed *co-translational translocation*
(figure 4.7). (It should not be confused with the post-translational
translocation that is associated with the integration of mitochondrial
and chloroplast proteins into those organelles.) Once in the lumen of
the rough ER, the newly synthesised protein would be inaccessible to
non-penetrating reagents, such as proteases.

　　Signal peptides are present not only in extracellular proteins of
eukaryotic cells, but also in proteins located in their ER, Golgi complex
and lysosomes, those at the external face of the plasma membrane, and
even those in the **periplasm** of prokaryotes. The *N*-terminal hydrophobic sequence is the characteristic that distinguishes such secretory
proteins from those of the cytosol, the nucleus, mitochondria or

Periplasm: The aqueous space
between the plasma membrane and
the cell wall of prokaryotes.

Pre-pro-albumin	M K W V T F L L L L F I S G S A F S	**Table 4.1**
Pre-IgG light chain	M D M R A P A Q I F G F L L L L F P G T R C	**Amino acid sequence of**
Pre-lysozyme	M R S L L I L V L C F L P L A A L G	**signal peptides**

In the *N*-terminal signal peptides of these secretory proteins, note the preponderance of hydrophobic amino acids (boxed). In each case, the cleavage site for signal peptidase comes after the last amino acid shown. (One-letter code for amino acids as in appendix A).

chloroplasts. The role of the signal peptide in translocation has been confirmed by recombinant DNA experiments. The nucleotide sequence corresponding to the signal peptide of **pre-pro-insulin** was fused to the gene for a cytosolic bacterial enzyme. When expressed *in vitro*, the chimaeric protein was incorporated into microsomal vesicles, in exactly the same manner as for a secretory protein. (In principle, hydrophobic stretches in nascent cytosolic proteins could interact with the secretory system. In practice, such proteins probably undergo co-translational folding that shields such non-polar regions from the components that carry out translocation.) In a few exceptional cases, such as ovalbumin, the signal peptide may be an *internal* sequence of hydrophobic amino acids close to the *N*-terminal (residues 21–47 for ovalbumin).

Pre-pro-insulin: The initial precursor of insulin, with its signal (*pre-*) sequence and its cleavable C-peptide (*pro-*) sequence.

4.2.2　Co-translational translocation involves a signal recognition particle and its receptor

What is the mechanism that directs the nascent secretory protein to the membrane of the rough ER? The answer to this question has come largely from studies of protein biosynthesis in the presence of microsomes that reconstituted this process *in vitro*.

Fractionation of the cytosolic components of the cell-free translation

Figure 4.7
Translocation of proteins synthesised on the rough ER

A ribosome (left) initiates translation of the mRNA for a secretory protein. The *N* terminal hydrophobic signal peptide (✺✺✺ N) passes through a pore in the membrane of the rough ER. In the lumen, the signal peptide is removed and the polypeptide undergoes folding. Release of the completed protein is followed by dissociation of the ribosomal subunits from the membrane and from the mRNA (right).

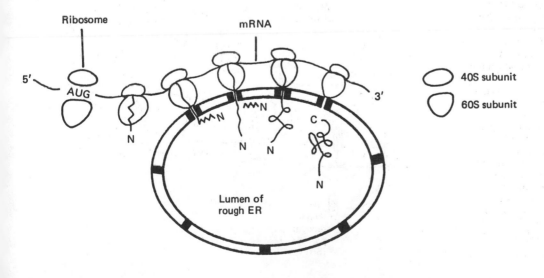

system led to the identification of the *signal recognition particle, SRP*. This large complex was susceptible to both proteases and ribonucleases, and was shown to contain 6 polypeptides and one copy of a specific 7S RNA of 300 nucleotides. The SRP binds to the growing polypeptide chain when it is about 70 amino acids long. This result is consistent with binding to the signal peptide: 50 or so amino acids must be linked together before the *N*-terminal emerges from the large ribosomal subunit. The component of the SRP involved has been identified by cross-linking studies: a photo-activatable derivative of lysine, incorporated into the signal peptide, labelled the 54 kDa polypeptide. Kinetic studies demonstrated that binding of SRP to the nascent polypeptide results in the temporary arrest of translation. This mechanism ensures that translocation is strictly coupled to chain elongation. The block is relieved by a protease-sensitive component in the microsomal membrane. Treatment of the microsomes with 0.5 M KCl led to the release of this membrane polypeptide into solution. This so-called *docking protein* (or SRP receptor) has been characterised as an integral protein of the rough ER, with an α subunit (69 kDa) and a β subunit (30 kDa). The cloned α subunit has an *N*-terminal membrane anchor, and a 52 kDa cytosolic domain with a positively-charged region that may interact with the 7S RNA. Binding of the SRP to the docking protein is associated with continued translation and release of the SRP. The events thought to be involved in co-translational translocation are summarised in figure 4.8.

The initial binding of the ribosome to the membrane of the rough ER, *via* the interaction of SRP and its receptor, is further stabilised by an uncharacterised protein-mediated process that involves the 60S ribosomal subunit. The details of the mechanism whereby the growing polypeptide chain crosses the membrane are as yet unclear. The

Figure 4.8
A possible mechanism for co-translational translocation

The signal peptide (**WWW N**) binds the signal recognition particle SRP), which then associates with the docking protein (SRP receptor) in the membrane of the rough ER. After release of the SRP, the signal peptide is thought to bind to a signal sequence receptor, a 35 kDa integral membrane glycoprotein. The creation of a transient pore allows cotranslational translocation. Signal peptidase cleaves the *N*-terminal hydrophobic sequence. (In a few cases, the signal peptide may not be removed; it remains as a membrane anchor for the protein concerned.) Continued translation generates a folded polypeptide, shown here to be anchored to the membrane by a nonpolar region (**—ƖƖƖƖƖ—**) near the *C*-terminal.

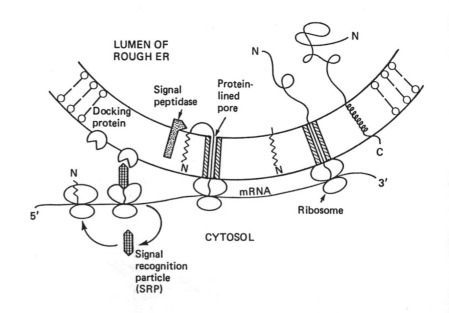

LUMEN OF ROUGH ER

Signal peptidase

Protein-lined pore

Docking protein

mRNA

Ribosome

CYTOSOL

Signal recognition particle (SRP)

existence of a transient protein-lined pore has been postulated, but not confirmed. Ribophorins I and II, two membrane proteins specific to the rough ER, may be involved.

4.2.3 Numerous post-translational modifications occur in the lumen of the rough ER

Nascent secretory proteins are vectorially discharged into the lumen of the rough ER. Here they encounter the first of many covalent changes in their structure: the signal peptide is removed. Not all such post-translational modifications occur in the rough ER; other important changes take place in the Golgi complex (section 5.2), extracellularly, and elsewhere. The consequences of these covalent alterations vary, depending on the protein concerned (table 4.2). The formation of disulphide bridges (probably catalysed by the enzyme, protein disulphide isomerase) between closely apposed cysteine residues may occur before translation is complete; we will not consider this process further. (We should not forget that many cytosolic proteins also undergo covalent modifications that influence their activity. The classic example is the control of glycogen metabolism by the hormone-induced phosphorylation of glycogen synthetase and phosphorylase.) We shall concentrate here on one system of particular interest: glycosylation of the G protein of vesicular stomatitis virus (VSV), a process that is initiated in the rough ER and completed in the Golgi complex.

VSV is an **enveloped RNA virus** that infects mammalian cells. During the infection, the host's translational machinery is usurped by the virus, which directs the synthesis of its own proteins, including the G protein. Multiple copies of the G protein are inserted into the membrane envelope, where they form 'spikes' on the viral surface that will mediate infection of another cell. Since it represents the vast majority of the glycoprotein synthesised by the infected cell, the G protein has become a widely used model system for the study of glycosylation in mammalian cells.

G protein is synthesised with a signal peptide on membrane-bound ribosomes, as described previously (section 4.2.2). As it enters the lumen of the rough ER, the hydrophobic pre-sequence is removed by a specific endoprotease, *signal peptidase* (figure 4.8). In microsomes prepared from dog pancreas, this activity is associated with an integral membrane complex. Of its 4–6 constituent subunits, with molecular masses in the

Enveloped virus: Virus surrounded by a membrane derived from the plasma membrane of the host cell.

Modification	Physiological role	Example	
Partial proteolysis	Activation of precursor	Pro-insulin	**Table 4.2**
Glycosylation	(Various)	Glycoproteins	**Post-translational processing**
Sugar phosphorylation	Intracellular targeting	Lysosomal enzymes	
Hydroxylation	Stabilisation of fibrous proteins	Collagen	
Fatty acylation	Membrane anchoring	Myelin basic protein	

range 12–25 kDa, two are glycoproteins and one has signal peptidase activity. The non-catalytic peptides may form part of the pore through which the secretory protein is thought to pass during translocation. Glycosylation of the nascent G protein takes place even before translation is complete. The process is referred to as N-*linked glycosylation*, since the sugar residues are attached to the amide nitrogen in the side-chain of an asparagine residue. (*O*-linked glycosylation of the hydroxyl-containing amino acids, serine and threonine, is a totally different process that occurs in the Golgi complex; section 5.2.2.) The signal for attachment of the oligosaccharide chain appears to be a simple tripeptide sequence: Asn.X.Ser/Thr, where the target asparagine is followed by any amino acid (X) and then by either of the hydroxyl-containing amino acids. For efficient glycosylation, the target sequence has to be in an exposed position on the folded polypeptide.

In what form are the sugar residues added to the nascent glycoprotein? Incubation of VSV-infected cells with radioactive sugars, followed by isolation and characterisation of the labelled species, has defined the nature of the sugar donor in some detail. The sugar residues are added *en bloc* in the form of an oligosaccharide chain covalently linked to a polyisoprenoid lipid, *dolichol* (figure 4.9). The dolichol-oligosaccharide is synthesised in the membrane of the rough ER by the sequential action of *glycosyl transferases* which have an exquisite specificity for both the donor sugar derivative and the acceptor structure (figure 4.10). The first seven sugars (those underlined in figure 4.10b) are added at the cytosolic face of the rough ER. An integral membrane protein is thought to translocate this dolichol derivative across the lipid bilayer. The remaining seven sugars are then added at the luminal face of the rough ER. The glucosylated dolichol-oligosaccharide is the substrate for an oligosaccharide transferase, which presumably recognises the target asparagine.

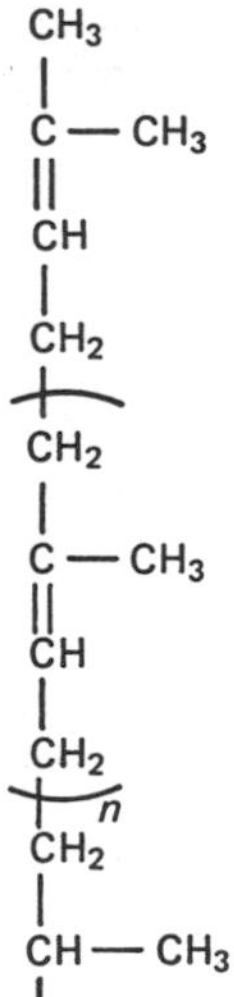

The glucose residues are quickly removed from the glycosylated G protein (figure 4.10b) by two specific glucosidases. One of them, glucosidase II, has been located by immuno-electron microscopy in transitional elements of the ER, close to the Golgi complex (as seen in figure 5.7). This association is significant. It is thought that transport vesicles bud off from the ER and ferry newly synthesised glycoproteins to the Golgi complex. This process has been studied in cell lines with a specific temperature-sensitive mutation. At the permissive temperature (32°C), G protein undergoes further processing by Golgi-associated enzymes in a reconstituted microsomal preparation. At the non-permissive temperature (40°C), this processing is blocked because the glycoprotein does not reach the Golgi complex. The transfer process involves two steps: a temperature-sensitive, ATP-dependent export (budding), and a temperature- and ATP-insensitive delivery (fusion). We shall soon learn the details of this process, and the identify of the cytosolic components involved.

We do not have the space here to consider the many other post-translational modifications that take place in the ER. The reader is referred to the reviews listed in section 4.7.

Figure 4.9
Structure of a dolichol

In mammals, *n* is generally 14–18; the degree of unsaturation is also variable. The hydroxyl group is phosphorylated in dolichol phosphate.

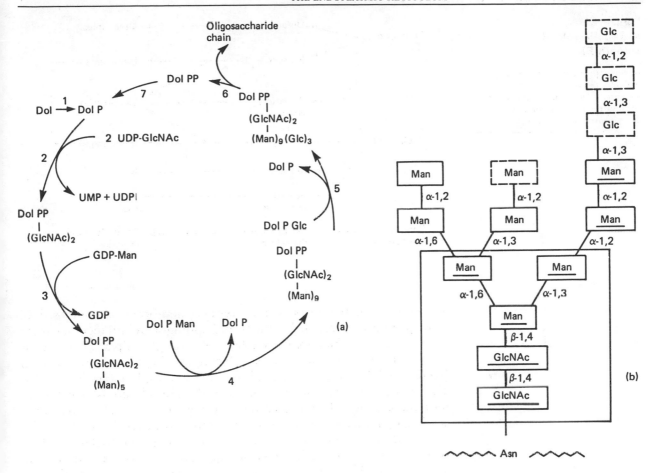

**Figure 4.10
Dolichol-mediated
glycosylation of proteins**

(a) The oligosaccharide chain is assembled on a dolichol pyrophosphate carrier (Dol PP), derived by phosphorylation of dolichol (Stage 1) and transfer of GlcNAc 1-P from UDP-GlcNAc (Stage 2). Sugars are added from nucleotide derivatives (Stage 3), or from DolP derivatives (Stages 4 and 5). The glucosylated oligosaccharide chain is transferred to the nascent glycoprotein (Stage 6). DolP is regenerated (Stage 7) by a specific phosphatase. (b) The asparagine (Asn)-linked oligosaccharide chain undergoes enzymic processing which removes those sugars indicated in the dashed boxes. The large box encloses the 'core-mannose' structure. (Sugar abbreviations are: GlcNAc, *N*-acetyl-D-glucosamine; Man, D-mannose; Glc, D-glucose).

4.3   The smooth endoplasmic reticulum

The smooth ER is particularly extensive in the parenchymal cells of the liver (the hepatocytes) and in those cells making steroid hormones. A common feature of these cells is a high rate of lipid biosynthesis. In fact, the smooth ER is active in all cells, where it makes the amphipathic lipids that will be incorporated into membranes (figure 4.11). In hepatocytes, the smooth ER is the major site for the detoxification of foreign compounds and other chemicals. In these cells and in those of the renal cortex, it is also the site for the release of free glucose (in the process of gluconeogenesis) by a membrane-bound glucose 6-phosphatase complex. In skeletal muscle cells, the smooth ER is specialised to form the *sarcoplasmic reticulum,* which actively sequesters Ca^{2+} ions and thereby regulates muscle contraction.

4.3.1   The smooth ER synthesises most of the cell's membrane lipids

In a growing cell, there is a continuous increase in the area of the plasma membrane and all the intracellular membranes. About

Figure 4.11
Metabolic role of the
smooth ER

In this representative diagram, not all
the activities of the smooth ER are
indicated. Also, some of the activities
are limited to certain organs (such as
detoxification only in the liver).

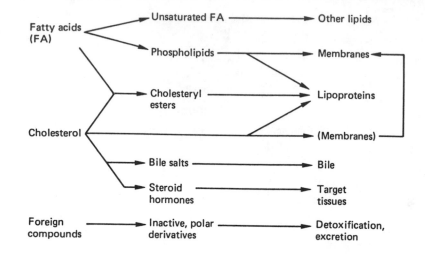

50 per cent of the mass of an average membrane is composed of
lipid, primarily phosphoglycerides and sphingomyelin (together these
comprise the phospholipids), cholesterol, and glycolipids. Fatty acids
are synthesised in the cytosol, glycolipids in the Golgi complex
(section 5.2.2), plant galactolipids in the chloroplast (section 9.4.3).
The smooth ER assembles most of the cell's phospholipids, which form
the bulk of the lipid bilayer (see, for example, table 1.5). (One exception
is diphosphatidylglycerol, or cardiolipin, synthesised by the outer
mitochondrial membrane.) Also, the later stages of cholesterol biosynthesis are catalysed by microsomal enzymes. We cannot go here
into the details of the biosynthesis of lipids by the smooth ER;
appropriate reviews are indicated in section 4.7. Nevertheless, we ought
to state here two key points concerning this process:

(a)   the reactions involved are catalysed by enzymes embedded in the
      smooth ER, but with their active site facing the cytosol;
(b)   nascent phospholipids are incorporated into the cytosolic half of
      the lipid bilayer.

We shall see the significance of these points later when we consider
membrane biogenesis (section 4.4).

$$A—H + O_2 + NADPH + H^+$$

Cytochrome P450-linked
mixed function oxidase

$$A—OH + H_2O + NADP^+$$

Figure 4.12
Activity of mixed-function
oxidases

In this reaction a hydrocarbon (such
as a steroid, A–H) undergoes
hydroxylation. One atom of the
molecular oxygen is incorporated into
the — OH group, the other into H_2O.
The mixed function oxidase, or mono-
oxygenase, is so called because it
utilises as substrates both an oxidising
agent (O_2) and a reducing agent
(NADPH).

4.3.2   An unique electron transport system catalyses hydroxylation reactions

There is one reaction common to the biosynthesis of cholesterol,
bile acids and steroid hormones, and to detoxification mechanisms:
hydroxylation. In hepatocytes, this type of reaction is catalysed by a
so-called *mixed-function oxidase* (figure 4.12), an integral membrane
protein of the smooth ER. There are many such enzymes, each specific
for one substrate (or a small group of related substrates). They

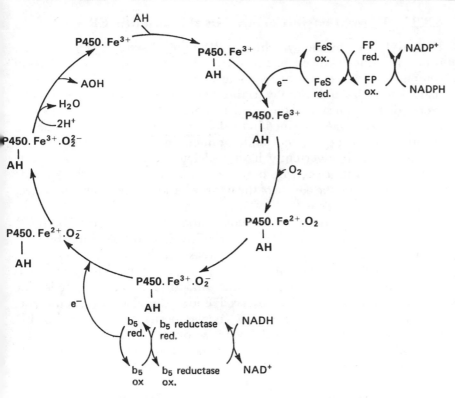

Figure 4.13

The hepatic cytochrome *P*450 system

This cyclic series of reactions catalyses the hydroxylation of a substrate (AH) such as an aryl hydrocarbon. The starting point is the top left-hand corner. Cytochrome *P*450 in the smooth ER can accept an electron from either of two electron transport systems, one linked to NADPH (top right) and one to NADH (bottom centre). The former contains a flavoprotein (FP) that acts as an iron-sulphur protein (FeS)–NADPH reductase. The latter contains cytochrome b_5 and its associated reductase. (The adrenal cytochrome *P*450 system differs in having a ferredoxin-like protein, adrenodoxin, in place of the FeS protein.) Oxidised and reduced states of each protein are abbreviated ox. and red., respectively.

are linked to an electron transport system that is characterised by **cytochrome *P*450** (figure 4.13). Although there are some superficial resemblances to the mitochondrial electron transport chain (section 8.2.1), the microsomal system is distinguished by the fact that it utilises NADPH as an electron donor, and it is not involved in the conservation of metabolic energy (ATP synthesis).

There are at least 40 structural variants of cytochrome *P*450. The genes for some of them have been cloned in the form of cDNA, as has that for cytochrome b_5. The deduced amino acid sequences show that each has a highly hydrophobic region which could act as a membrane anchor; this region is at the *N*-terminal for *P*450, but near the *C*-terminal for b_5. The *P*450 cytochromes are synthesised by membrane-bound ribosomes as described for secretory proteins (section 4.2.1), whereas cytochrome b_5 is synthesised on cytosolic ribosomes. Insertion of b_5 into the lipid bilayer therefore involves a post-translational mechanism. We do not presently understand how such a protein becomes incorporated specifically into the microsomal membrane. The recombinant DNA studies further indicate that there are several families of functionally related cytochrome *P*450 genes (and pseudo-genes). On mouse chromosome 9 the seven exons of the $P_3$450 gene must be separated by substantial introns, for the mature mRNA (2.1 kb) is less than a third of the length of the gene itself.

Cytochrome *P*450: Family of structurally related cytochromes located (in hepatocytes) in the ER (*P* refers to its *p*articulate, or microsomal, localisation; 450 refers to the wavelength, 450 nm, of maximum light absorption by its complex with carbon monoxide).

4.3.3   Detoxification occurs in the smooth ER

Hydroxylation (by the cytochrome $P450$-linked system; section 4.3.2) is one reaction involved in the metabolism and excretion of foreign compounds and certain endogenous chemicals. The rationale here is that such compounds are invariably hydrophobic; before they can be excreted, they must be converted to polar derivatives (figure 4.14). For example, *bilirubin*, formed in the liver from the breakdown of haem, is converted to the diglucuronide adduct. In some cases the Phase I reactions may inadvertently have a deleterious effect. Polyaromatic hydrocarbons induce cancers because they are converted to diolepoxide derivatives during the course of the hydroxylation stage; such derivatives are potent carcinogens.

One striking feature of the detoxification process in the smooth ER is its *inducibility*. For example, administration of phenobarbital, a barbiturate drug, causes a 5-fold increase in the synthesis of certain microsomal mixed-function oxidases in hepatocytes. This increase is associated with proliferation of the smooth ER in these cells, the outcome of which can be observed under the electron microscope. Withdrawal of the drug results in the destruction of excess ER membrane by autophagic breakdown in secondary lysosomes (figure 6.8). The molecular mechanism of this induction process is not fully understood.

4.3.4   The smooth ER sequesters calcium ions

In many cell types, the smooth ER plays an important role in the regulation of calcium (Ca^{2+}) levels within the cell. Normally, the concentration of free Ca^{2+} in the cytosol is maintained at a low level, around 10^{-7} M. An increase in free Ca^{2+} to micromolar levels triggers several effects, which are mediated by calcium-binding proteins, such as *calmodulin*. Two systems have been particularly well studied. Firstly, **Berridge** has provided convincing evidence that the smooth ER is involved in the mode of action of those peptide hormones which use

Figure 4.14
Detoxification reactions in the smooth ER

Phase I brings about hydroxylation of the substrate (S). Phase II further increases polarity by conjugation with UDP–glucuronic acid (UDP–GlcUA) or 3′-phosphoadenosyl 5′-phosphosulphate (PAPS), in reactions catalysed by the appropriate specific transferase. The glucuronide and sulphate derivatives are water-soluble, and can be excreted in the urine.

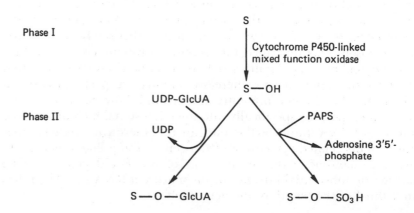

inositol trisphosphate (IP_3) as a second messenger. Secondly, **Carafoli** has described in detail the role of the specialised internal membranes (the *sarcoplasmic reticulum*) of skeletal muscle cells in regulating contraction.

Phosphoinositides (figure 4.15) are quantitatively minor phospholipid components present in the plasma membrane of most cell types. Their biosynthesis and degradation can be studied by incubating cells with radioactive [^{32}P]-phosphate; the labelled phosphorylated lipids can be subsequently isolated by chromatography. From such studies, it was found that stimulation of pancreatic cells by acetylcholine led to a rapid increase in the incorporation of label into phosphatidylinositol (PI). This so-called '*PI response*' was found to be widely distributed amongst animal tissues (table 4.3). In platelets pre-labelled with [^{32}P]-phosphate, exposure to thrombin caused a rapid loss of label from PI 4,5-bisphosphate, and its appearance in the water-soluble compound 1,4,5-inositol trisphosphate (IP_3). The proposed sequence of events is depicted in figure 4.16. Experimental evidence supporting a role for the smooth ER in this process came from studies in which rat pancreatic cells were permeabilised by incubation in a medium of low Ca^{2+} concentration. Treatment of such cells with sub-micromolar concentrations of IP_3 induced the release of intracellular Ca^{2+}, as measured by a calcium electrode. After cell fractionation, IP_3 liberated Ca^{2+} only from a microsomal fraction, not from mitochondria or secretory granules.

A similar role in sequestering Ca^{2+} has been established for the specialised internal membranes of muscle cells, the *sarcoplasmic reticulum*. This extensive system of interconnected tubules ramifies between the contractile elements (figure 4.17). A membrane-bound Ca^{2+}-ATPase pumps Ca^{2+} ions from the cytosol into the lumen of the sarcoplasmic reticulum, thereby generating a massive transmembrane gradient (approximately 10^{-7} M Ca^{2+} in the cytosol, 10^{-3} M in the lumen). Contraction of the muscle fibre is initiated by the depolarisation brought about by an action potential arriving at the neuromuscular junction. As a result, Ca^{2+} is liberated from the sarcoplasmic reticulum, and the Ca^{2+}-dependent contractile process is stimulated. IP_3 has been shown to mediate Ca^{2+} release in iris muscle, but it is not known whether it has a similar role in other muscle types.

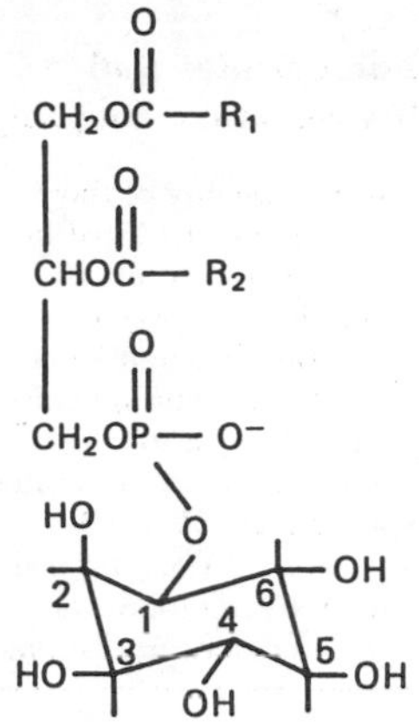

**Figure 4.15
Structure of
phosphoinositides**

Phosphoinositides are phosphate esters of phosphatidylinositol, PI (shown here). In the diacylglycerol backbone, the fatty acyl chains (R_1 and R_2) may differ in length or degree of unsaturation. Phosphate may be esterified to the hydroxyl group on C4 of the *myo*-inositol ring (to give PI 4-phosphate), or on both C4 and C5 (to give PI 4,5-bisphosphate).

Stimulus	Responsive tissue(s)
Thrombin	Platelets
Antigen	Mast cells
Glucose	Pancreatic B cells
Acetylcholine	Various secretory tissues, smooth muscle

**Table 4.3
Examples of the 'PI response'**

Figure 4.16
Phosphoinositides and
stimulus-response coupling

The ligand may be any of those
stimulatory components listed in
table 4.3. The ligand-receptor
complex stimulates a specific
phosphodiesterase in the plasma
membrane. This enzyme splits PI 4,5-
bisphosphate (PIP_2) into
diacylglycerol (DAG) and inositol
1,4,5-trisphosphate (IP_3). IP_3
stimulates release of Ca^{2+} from the
endoplasmic reticulum into the
cytosol. ($\oplus$ indicates stimulation.)

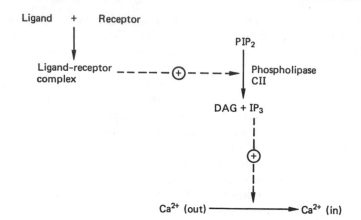

4.4 Membrane biogenesis in the endoplasmic reticulum

We saw earlier how the ER is the site of biosynthesis of many of the
phospholipids (section 4.3.1) and proteins (section 4.2.1) that are
incorporated into the cell's membranes. In considering these processes
further, we need to bear in mind some of the key features of membrane
structure and function that we touched on in section 1.3.1: biological

Figure 4.17
Sarcoplasmic reticulum of
vertebrate skeletal muscle

Bundles of myofibrils (c) are
separated by elements of the
sarcoplasmic reticulum (a), a
membranous network closely
associated with the T-tubules (b).
The sarcoplasmic reticulum releases
and sequesters the calcium ions
necessary for muscular contraction.
[Electron micrograph taken from
Macleod, A. G. (1975). *Cytology*, The
Upjohn Company, Kalamazoo;
reproduced by permission of the
publisher.]

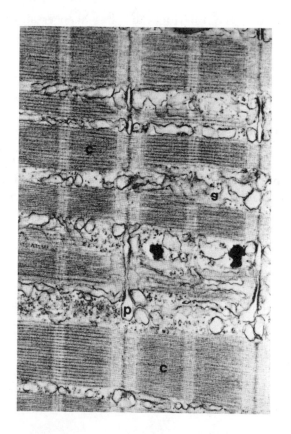

membranes are asymmetric, fluid, lipid bilayers containing oriented globular proteins with specific functions. How are such structures assembled in the ER and distributed to the appropriate intracellular site?

Phospholipids, in the course of their synthesis by membrane-bound enzymes of the ER, become incorporated into the inner leaflet of the lipid bilayer. It follows that there must be mechanisms for: (a) distributing newly synthesised phospholipids to the outer leaflet; (b) achieving the asymmetric lipid composition of the bilayer (for example, with phosphatidylethanolamine relatively enriched in the inner leaflet); (c) incorporating those membrane lipids not synthesised in the ER (such as gangliosides). These processes are mediated largely by **lipid transfer proteins**. In each case, the transfer protein acts specifically on one lipid 'substrate', but we do not yet have a clear picture of the mechanism of action involved.

Lipid transfer proteins: There appear to be two classes of these proteins: membrane-bound proteins which equilibrate newly synthesised lipids between the two monolayers of the ER membrane; and soluble proteins which transport newly synthesised lipids to mitochondria and other organelles.

Glycolipids may be assembled in the specific organelle where they are localised (as with chloroplast galactolipids), or in the Golgi complex for distribution to other intracellular sites (mainly the plasma membrane; section 5.2.2). In the case of the Golgi-assembled glycolipids, the sugar transferases responsible for synthesising the oligosaccharide side-chains are themselves located in the lumen of that organelle. Hence, the asymmetric distribution of membrane glycolipids is a consequence of their biosynthesis: their oligosaccharide portions always face away from the cytosol.

A corresponding biosynthetic mechanism achieves the asymmetric distribution of membrane proteins: polypeptides translated on membrane-bound ribosomes are inserted asymmetrically into the lipid bilayer of the rough ER (section 4.2.1). However, we also need to consider how such proteins reach their correct intracellular site, and how the appropriate portions or *domains* of the protein are disposed on either side of the membrane.

Blobel has extended his signal-peptide hypothesis (section 4.2.1) to consider in more general terms how membrane proteins become anchored in a lipid bilayer. He proposes the existence of '*stop-transfer*' *sequences* of amino acids (figure 4.18). There are many well-studied proteins which show such features. For example, **glycophorin** has the distribution shown in figure 4.19a. As expected, the oligosaccharide chains of this glycoprotein have an extracellular disposition. In the case of the cholinergic *acetylcholine receptor,* a multimeric protein of the synaptic membrane of certain neurones, multiple hydrophobic segments cause the polypeptide to loop back and forth across the lipid bilayer (figure 4.19b). The association of five amphiphilic helices in this protein may generate an aqueous channel or pore (diameter 0.6 nm) traversing the membrane. Amphiphilic helices are also thought to operate in the targeting of mitochondrial proteins (figure 8.18).

Glycophorin: Glycoprotein present in the plasma membrane of mammalian erythrocytes.

A hydrophobic segment of the polypeptide chain is not the only means of anchoring an integral membrane protein. Increasingly, examples are being described where a covalently bound *lipid* moiety

Figure 4.18
Models for anchoring membrane proteins

According to Blobel, 'stop-transfer' sequences orientate proteins in biological membranes. During translation, the hydrophobic 'stop' sequences (shown here by the helical lines) anchor such proteins in the lipid bilayer; the hydrophilic 'transfer' sequences are excluded from the membrane. (a) In the simplest case, the protein is orientated with its *N*-terminal domain in the non-cytosolic medium (OUT), and its C-terminal domain in the cytosol (IN). (b) With two 'stop' sequences, both domains face out; an internal domain is cytosolic. (c) With a highly hydrophobic signal peptide, cessation of translocation may generate an extensive *C*-terminal domain either cytosolic (as shown here) or extracellular. (d) Multiple 'stop' sequences may result in a complex disposition of domains.

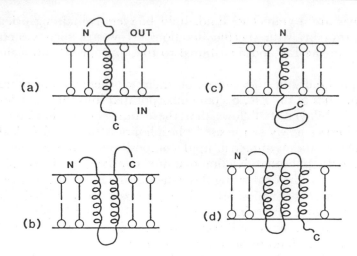

Trypanosomes: Protozoal parasites that cause sleeping sickness in humans.

achieves the same end. For example, when mammalian cells infected with VSV were incubated with radioactive palmitic acid (C16:0), labelled fatty acid was incorporated into the G protein of the viral envelope. Isolation of the radioactive protein, followed by peptide mapping, pinpointed the fatty acid: it was covalently linked *via* a thioester bond to a cysteine residue located in the cytosolic domain. Presumably the lipid stabilises the protein in the plasma membrane. (Note, however, that this fatty acylation step probably occurs in the Golgi complex.) A similar feature is present in the *variant surface glycoprotein* (VSG), a protein of the plasma membrane of certain **trypanosomes.** The newly synthesised VSG undergoes a curious endoproteolytic cleavage that removes a hydrophobic *C*-terminal membrane anchor. In its place, a residue of phosphatidylinositol (bearing two myristyl (C14:0) residues) is linked covalently to the polypeptide chain *via* a phosphorylated oligosaccharide chain. Even this novel membrane anchor can be removed. Cleavage by a specific phospholipase releases the VSG, a process that allows the trypanosome to change its surface coat or even to lose it altogether.

The targeting of membrane proteins is currently an active area of research. In the case of polypeptides synthesised on cytosolic ribosomes, the nascent proteins are thought to bear recognition sequences which interact with receptor proteins of the appropriate organelle. Membrane proteins of the mitochondrion (section 8.4.1) and of the chloroplast (section 9.5.4) are thought to be targeted in this way. Subsequent enzyme-mediated removal of the recognition sequence would prevent loss of the protein from the organelle. Proteins translated on membrane-bound ribosomes have several possible destinations: the lipid bilayer of the ER itself, of the Golgi complex, of lysosomes and other vesicles, or of the plasma membrane. Once again, the existence of recognition sequences and specific receptors are postulated, but these are poorly documented. (Proteins such as the G protein of VSV may reach the

(a)

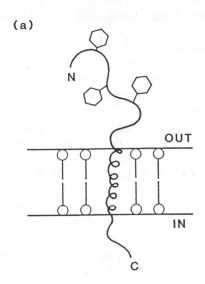

(b)

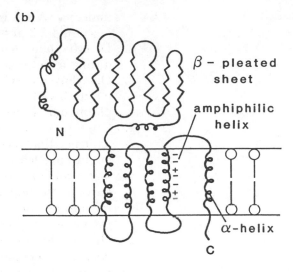

**Figure 4.19
Orientation of membrane proteins**

(a) Glycophorin has an α-helical sequence (residues 73–91) of hydrophobic amino acids that anchors the protein in the erythrocyte membrane. The extracellular (OUT) *N*-terminal and cytosolic (IN) *C*-terminal domains have a preponderance of polar residues. Charged amino acids mark the boundary between the membrane and the aqueous medium. The *N*-linked carbohydrate residues (⬡) face out. (b) The subunits of the cholinergic acetylcholine receptor are thought to have multiple transmembrane α-helical segments. In one such segment, every fourth residue is polar (charged); this amino acid sequence generates an amphiphilic helix, with a hydrophilic face and a hydrophobic face.

plasma membrane because they *fail* to meet a corresponding receptor *en route*). How each receptor protein becomes localised in a given membrane is also a matter for speculation.

4.5 Summary

The endoplasmic reticulum (ER) is the major site for the biosynthesis of the lipid and protein components of cellular membranes. It forms an extensive array of cisternae and tubules that ramifies throughout the cytosol. It is morphologically and functionally specialised in the form of rough ER, with bound ribosomes, and smooth ER, lacking ribosomes. The rough ER synthesises secretory proteins. Its membrane contains a 'docking protein' and other uncharacterised proteins, which interact with the signal recognition particle and the signal peptide of the nascent polypeptide, so as to bind the associated ribosome. Within the lumen of the rough ER, the signal peptide is removed and the first stages of post-translation modification occur. Transport vesicles ferry secretory proteins from the ER to the Golgi complex. The smooth ER is the site of many biosynthetic processes, in particular those associated with lipid metabolism and with detoxification. Some of the reactions involved require the participation of an inducible electron transport system, characterised by the presence of cytochromes *P*450. The ER is enlarged by the ordered incorporation of newly synthesised lipids and specific proteins. It is thus the chief site of membrane biogenesis within the cell.

4.6 Study questions

1. List the post-translational modifications that secretory proteins undergo. What is the functional significance of such changes? (section 4.2.3 and table 4.2)

2. Describe how structural and functional asymmetry is maintained during membrane growth. (section 4.4)

3. Diacylglycerol stimulates a membrane-bound protein kinase. How might this effect be related to stimulus-response coupling involving phosphoinositides? (figure 4.16)

4. Some secretory proteins, such as ovalbumin, contain a signal sequence that is *not* cleaved by signal peptidase. What experiment might have led to this conclusion? (section 4.2.3)

5. The proliferation of the smooth ER that is induced by pheno-barbital (section 4.3.3), could result from the acquisition of 'old' membrane from other intracellular sites. How might you show that the induction process involves synthesis of 'new' membrane?

6. How could you demonstrate that signal peptidase complex and docking protein are necessary and sufficient to translocate nascent pre-proteins across microsomal membranes? (sections 4.2.1 and 4.2.2)

7. How might you determine experimentally the intracellular site where palmitic acid is esterified to the VSV G protein? (sections 4.4 and 4.2.3)

4.7   Further reading

Lipid biosynthesis

Coleman, R. and Bell, R. M. (1978). *J. Cell Biol.*, **76**, 245–253.
   (Asymmetric phospholipid synthesis by the smooth ER)
Bloch, K. (1965). *Science*, **150**, 19–28.
   (Cholesterol synthesis)

Translocation of membrane proteins

Blobel, G. and Dobberstein, B. (1975). *J. Cell Biol.*, **67**, 835–253.
   (The signal peptide hypothesis)
Evans, E. A., Gilmore, R. and Blobel, G. (1986). *Proc. Nat. Acad. Sci. USA*, **83**, 581–585.
   (The signal peptidase complex)
Walter, P., Gilmore, R. and Blobel, G. (1984). *Cell*, **38**, 5–8.
   (Protein translocation across the ER)

Post-translational processing

Eyre, D. R. (1980). *Science*, **207**, 1315–1322.
   (Biosynthesis of collagen)
Kornfeld, R. and Kornfeld, S. (1985). *Ann. Rev. Biochem.*, **54**, 631–664.
   (*N*-linked glycosylation of proteins)
Sefton, B. M. and Buss, J. E. (1987). *J. Cell Biol.*, **104**, 1449–1453.
   (Fatty acylation of proteins)

Membrane function (selected aspects)

Rothman, J. E. and Lenard, J. (1977). *Science,* **195**, 743–753.
  (The significance of membrane asymmetry)
Berridge, M. J. (1987). *Ann. Rev. Biochem.,* **56**, 159–193.
  (Phosphoinositides and stimulus-response coupling)
Nebert, D. W. and Gonzalez, F. J. (1987). *Ann. Rev. Biochem.,* **56**,
  945–993.
  (The cytochrome *P*450 super-family)

5 THE GOLGI COMPLEX

5.1 Introduction

We saw earlier (section 1.1.3) how **Golgi** in 1898 first described an 'internal reticular apparatus' that he observed by light microscopy of silver-stained neuronal cells. In fact, for many years the existence of this structure was doubted: some workers thought that it was an artefact generated by the staining procedure. Only in recent years has the real significance of the **Golgi complex** come to be appreciated. This change in attitude has come about partly as a result of electron microscopy, which has demonstrated convincingly that this organelle is present in almost all eukaryotic cells; and partly from biochemical studies (sections 5.2 and 5.3) that have revealed its important metabolic activity. As indicated in table 1.1, the cell's Golgi complex is likely to represent a single organelle made up of a network of dispersed but interconnected lamellar structures. Such an arrangement has been confirmed by high-voltage electron microscopy of thick (0.5 μm) sections of tissues impregnated with heavy metals (figure 5.1). In animal cells, the Golgi complex is usually located close to the nucleus in the region of the centrioles.

Golgi complex: Synonyms for this organelle include Golgi body, Golgi apparatus, and dictyosome. The term 'Golgi complex' is preferred in this text.

Under the electron microscope, the organelle presents a complex morphology (figures 5.2 and 5.3). The characteristic appearance is that of a curved stack of 4–6 parallel cisternae (or saccules), each about 1 μm in diameter, with the convex face closer to the nucleus. Associated with this structure are tubules and vesicles at either face, and vesicles (about 50 nm diameter) along the lateral surfaces of the parallel cisternae but never between adjacent cisternae. The components of the organelle appear to be polarised. The membrane of the cisternae gradually thickens as one progresses from the convex (*cis*) face to the concave (*trans*) face. The cholesterol concentration of the Golgi membranes increases in the direction *cis* to *trans*. Also, osmium preferentially stains the cisternae at the *cis* face (figure 5.4a). In secretory cells the orientation of the Golgi complex is particularly obvious (figure 5.5). The *cis* (or proximal) face is associated with transitional elements of the RER, from which small smooth-surfaced vesicles can be seen budding (a process more apparent in figure 5.7).

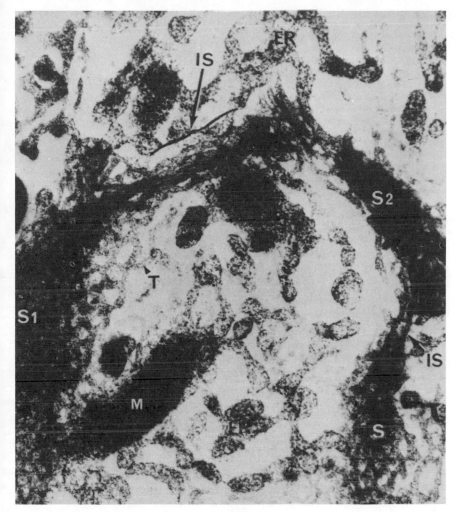

Figure 5.1
Connections within the Golgi complex

In this electron micrograph of a thick section of a Sertoli (testis) cell, three stacks (S, S1, S2) of Golgi cisternae are interconnected by tubular elements (IS). Golgi stacks are seen in face view (S1) or lateral view (S2). Other structures present are mitochondria (M), endoplasmic reticulum (ER), and tubules (T) associated with the concave face of the Golgi stack. × 32 000. [Taken from Rambourg, A. *et al.* (1979). *Amer. J. Anat.*, **154**, 455–476; reproduced by permission of the publisher, Alan R. Liss Inc.]

At the *trans* (or distal) face is a reticulum of interconnected tubules and associated vesicles: the so-called *trans*-Golgi network (visible in figures 5.1 and 5.2). Close by are condensing vacuoles which give rise to secretory vacuoles (or granules) containing the stored product (hormone, enzyme, etc.) ready for release by exocytosis.

The structural polarisation of the Golgi complex suggested by conventional electron microscopy has been confirmed by histochemical studies. Thiamine pyrophosphatase is preferentially associated in most cells with the *trans* cisternae (figure 5.4b). Acid phosphatase, the marker enzyme for lysosomes (see figure 6.1), is concentrated in elements of the *trans*-Golgi network. The location of other enzymes has been determined by immuno-electron microscopy (section 2.1.2), with the second antibody labelled with either peroxidase or colloidal gold. For example, galactosyl transferase (long used as a marker enzyme in centrifugation studies: section 2.2.1) is also located in the *trans* cisternae.

Biochemical studies have suggested that the Golgi complex may comprise three distinct sub-compartments (figure 5.6). After isopycnic

Figure 5.2
Elements of the Golgi complex

In this lateral view of a Golgi stack of cisternae, taken by electron microscopy of a Sertoli (testis) cell, five saccules (S) are evident. The most proximal *cis* saccule (CE) contains circular membranous profiles and faces a mitochondrion (M) and elements of the endoplasmic reticulum (ER). Below the most distal *trans* saccule is a network of tubules (TTN), the *trans*-Golgi network. Numerous small vesicles are associated with the lateral margins of the cisternae. × 60 000. [Taken from Rambourg, A. *et al.* (1979). *Amer. J. Anat.*, **154**, 455–476; reproduced by permission of the publisher, Alan R. Liss Inc.]

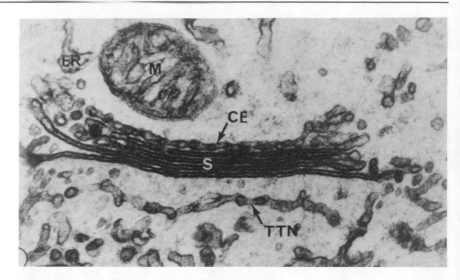

centrifugation, the most dense Golgi-derived vesicles appeared to correspond to the *cis* cisternae, whereas the least dense vesicles corresponded to the *trans* cisternae. The vesicles of intermediate density represented what has been termed the 'medial' cisternae. These differences in composition closely reflect the organisation of metabolic activity within the Golgi complex, as far as glycoprotein processing is concerned, at least (section 5.2). The enrichment of glycosyl transferase enzymes in Golgi fractions isolated by centrifugation, was compatible with other biochemical studies, which looked at the incorporation of radioactive sugars into glycoproteins by secretory cells *in vitro*. When examined by autoradiography, label (from [14C]-fucose or [3H]-galactose) appeared first over the Golgi complex. Here was experimental evidence that the organelle was involved in glycosylation, a process initiated in the endoplasmic reticulum (section 4.2.3). Recall that

Figure 5.3
Face view of the Golgi complex

In this freeze-fracture electron micrograph of a Golgi complex from a guinea-pig spermatocyte, cross fractures of the expanded peripheral portions of cisternae are indicated by arrows. Face views of some cisternae show a regular pattern of dimples and pore-like structures (fenestrations). [Taken from Fawcett, D. W. (1981). *The Cell*, 2nd edn, W. B. Saunders Co., Philadelphia; reproduced by permission of the publisher. Original micrograph courtesy of D. W. Fawcett.]

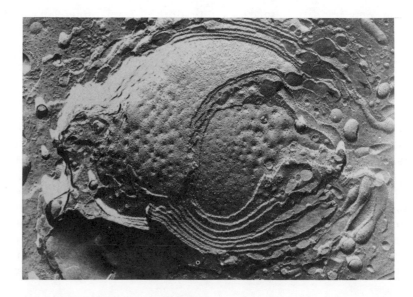

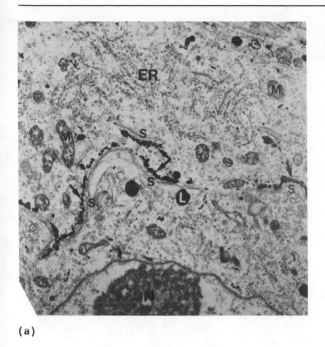

(a)

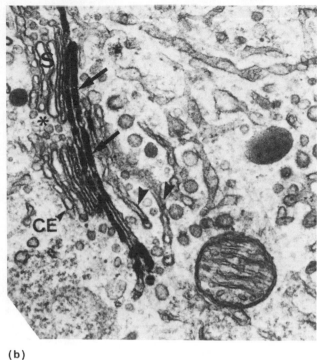

(b)

for *N*-linked glycoproteins, the oligosaccharide chains are transferred *en bloc* from a dolichol carrier to specific asparagine residues (figure 4.10). Enzymes in the ER rapidly remove the terminal glucose residues and one mannose, before the nascent glycoproteins are transferred to the Golgi complex *via* transport vesicles (figure 5.7), a process that takes about 10 minutes. Let us now follow the fate of these glycoproteins until they reach their final destination.

5.2 Metabolic processing in the Golgi complex

5.2.1 Oligosaccharide chains of glycoproteins undergo 'trimming' and further glycosylation

Mature glycoproteins can be classified into two groups depending on the composition of their oligosaccharide chains. The '*high-mannose*' type undergo little or no processing of their carbohydrate chains, which consist of 5–8 mannoses attached to the two internal *N*-acetylglucosamine residues. The *complex* type, however, are considerably modified: mannose residues are removed ('trimmed') and substituted with *N*-acetyl-glucosamine, galactose, sialic acid (and occasionally fucose) in a strict sequence. This processing, which takes place in the Golgi complex, is described in detail in figure 5.8. (Note that some glycoproteins may contain both 'complex' and 'high-mannose' oligosaccharide chains.)

This glycosylation pathway is determined by the substrate specificity of the glycosidases and glycosyl transferases involved. These enzymes

Figure 5.4
Polarisation of the Golgi complex

(a) Osmium preferentially stains the *cis* cisternae (arrows) at the convex face of the Golgi complex (s). (b) Thiamine pyrophosphatase, located with a histochemical stain, is found preferentially in the *trans* cisternae at the concave face (arrows). Both electron micrographs show a thin section through mammalian epididymis.
[Taken from Rambourg, A. and Clermont, Y. (1986). *Amer. J. Anat.*, 175, 393–409; reproduced by permission of the publisher, Allan R. Liss Inc.]

Figure 5.5
Golgi complex of a secretory cell

In this electron micrograph of a thin section of Brunner's gland (mouse duodenum), secretory granules appear as large dark vesicles at the *trans* face of the Golgi complex. N, nucleus; RER, rough endoplasmic reticulum. [Taken from Fawcett, D. W. (1981). *The Cell,* 2nd edn, W. B. Saunders Co., Philadelphia. Original micrograph courtesy of D. S. Friend.]

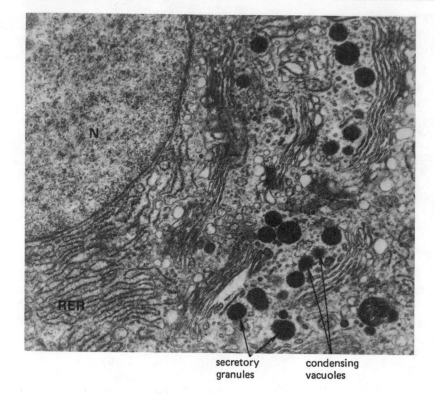

secretory granules   condensing vacuoles

Figure 5.6
Separation of Golgi sub-compartments

The components of a microsomal fraction from rat liver were separated on the basic of buoyant density by centrifugation through a sucrose gradient. Fractions from the gradient were assayed (diagram) for galactosyl transferase ($\bullet$), *N*-acetylglucosaminyl transferase ($\circ$), and *N*-acetylglucosamine 1-phosphate transferase ($\blacktriangle$). Three populations of Golgi-derived vesicles could be differentiated on the basis of their density and enzymic composition.

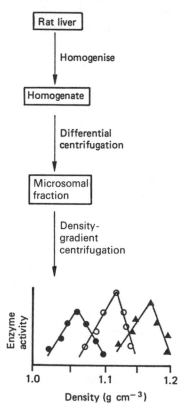

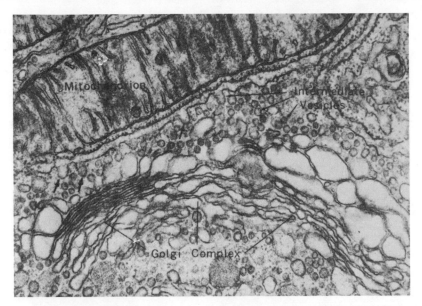

Figure 5.7
Vesicles link the endoplasmic reticulum with the Golgi complex

Intermediate (or transport) vesicles bud from the transitional ER (small arrows) and carry secretory proteins to the *cis* face of the Golgi complex. (Electron micrograph of a thin section of duodenal Brunner's gland.) [Taken from Fawcett, D. W. (1981). *The Cell*, 2nd edn, W. B. Saunders Co., Philadelphia. Original micrograph courtesy of D. S. Friend.]

are exquisitely specific, not only for the sugar involved, but also for the composition of the oligosaccharide chain which bears that terminal sugar. Furthermore, the sugar donors that act as substrates for the glycosyl transferases are nucleotide derivatives, synthesised in the cytosol. How do these highly polar compounds gain access to the lumen of the Golgi complex? Antiport proteins in the Golgi membrane exchange nucleotide-sugar for the corresponding nucleoside monophosphate (figure 5.9).

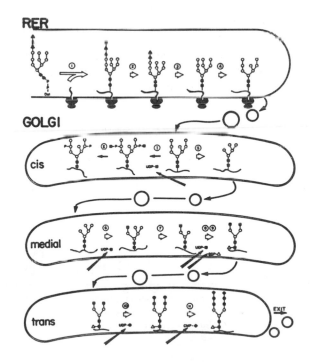

Figure 5.8
Processing of glycoproteins by the Golgi complex

Shown here is the modification of an *N*-linked oligosaccharide chain of a 'complex'-type glycoprotein. Each step is catalysed by a specific glycosyl transferase or glycosidase, acting invariably at the non-reducing end of the oligosaccharide chain and localised in a specific sub-compartment of the Golgi complex. (Symbols for sugars are: ■, *N*-acetyl-D-glucosamine; ○, D-mannose; ●, D-galactose; ◇, sialic acid (usually *N*-acetylneuraminic acid); △, L-fucose; ▲, D-glucose; DolPP, dolichol pyrophosphate). [Taken from Kornfeld, R. and Kornfeld, S. (1985). *Ann. Rev. Biochem.*, **54**, 631–664; reproduced by copyright permission of Annual Reviews Inc.]

Figure 5.9
Uptake of nucleotide-sugars by the Golgi complex

Reactions 1, 2 and 3 are catalysed by the appropriate glycosyl transferases. Reaction 4 is catalysed by nucleoside diphosphatase (probably the identical enzyme to thiamine pyrophosphatase, figure 5.4b). Specific antiport proteins (permeases) in the Golgi membrane are represented by circles. [Sugar abbreviations: GlcNAc, N-acetyl-D-glucosamine; Man, D-mannose; Gal, D-galactose; SA, sialic acid.]

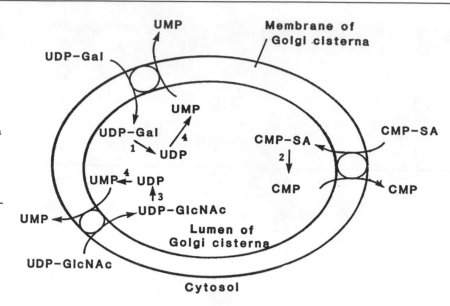

In a particularly interesting and well-studied case, lysosomal enzymes have been found to undergo a novel modification (described in more detail in figure 6.11 and section 6.4.2). The N-acetylglucosamine 1-phosphate transferase which initiates this process has been identified in the *cis* sub-compartment of the Golgi complex (figure 5.6). The net result of the modification is that nascent lysosomal enzymes are 'tagged' with a marker that acts as a signal for their incorporation into lysosomes. Here is an example of the intracellular fate of a newly synthesised protein being determined by a characteristic structural feature. It seems reasonable to assume that this model will be extended to account for the intracellular sorting of the proteins of other organelles also.

The precise location within the Golgi complex of the enzymes involved in a glycoprotein processing has been determined by biochemical studies (figure 5.6) and by immuno-electron microscopy. The results are summarised in table 5.1. Since there are apparently no direct

Table 5.1
Biochemical specialisation within the Golgi complex

Cis	Medial	*Trans*
N-acetylglucosamine 1-P transferase	N-acetylglucosaminyl transferases I + II	Galactosyl transferase
Protein–palmitoyl transferase	Mannosidases I + II	Sialyl transferase
	NADPH phosphatase	Fucosyl transferase
		Thiamine pyrophosphatase

The definition of *cis*, medial, and *trans* sub-compartments is given in section 5.1. Most of the enzymes are concerned with glycoprotein processing; their functions are outlined in figure 5.8. It is not known how each enzyme comes to be localised in a given sub-compartment.

connections between the cisternae of a Golgi stack, how do glycoproteins undergoing processing migrate from one saccule to another? We shall see later (section 5.3.2) that small vesicles probably transfer material between cisternae by a process of budding and fusion. An alternative hypothesis proposed that cisternae migrate through the Golgi stack; the most proximal *cis* saccule is re-formed by fusion of vesicles derived from the ER, and the most distal *trans* saccule would give rise to secretory and other vesicles. However, this hypothesis cannot account satisfactorily for the biochemical specialisation of the Golgi complex (table 5.1).

5.2.2 Other biosynthetic reactions take place in the Golgi complex

The processing of N-linked oligosaccharide chains is by no means the only metabolic event which takes place in the Golgi complex. This organelle carries out many other biosynthetic reactions, a considerable number of which relate in some way to glycosylation or the post-translational modification of secretory proteins.

(a) *Fatty acylation:* we described this covalent modification of membrane proteins in section 4.4.

(b) *O-linked glycosylation*: carbohydrate chains can also be attached to the hydroxyl-containing amino acids (serine, threonine and hydroxylysine) of some glycoproteins. This process of O-linked glycosylation is quite different from that of N-linked glycosylation described earlier (figures 4.10 and 5.8). The oligosaccharide chains are not assembled on a dolichol carrier. Instead, specific glycosyl transferases add the appropriate sugar (from the nucleotide–sugar derivative) to the non-reducing terminal of a growing oligosaccharide chain. The carbohydrate–protein link is invariably either D-xylose to serine (or threonine), or N-acetyl-D-galactosamine to serine (or threonine).

(c) *Collagen biosynthesis:* in the special case of collagen, disaccharide units (D-glucose-D-galactose-) are coupled to some hydroxylysine residues by specific galactosyl and glucosyl transferases. The procollagen chains undergo alignment to form a triple helical structure, as well as the formation of inter-chain disulphide bonds, within the Golgi complex.

(d) *Glycolipid biosynthesis:* assembly of cerebrosides, gangliosides, and other glycolipids (figure 5.10) that probably occurs in an analogous manner to that of O-linked glycosylation. The reactions of both processes are probably catalysed by the same glycosyl transferases.

(e) *Sulphation:* the tyrosine residues of some proteins, the galactose moiety of galactosylceramide (galactocerebroside), and some hexosamine residues of proteoglycans, undergo sulphation catalysed by different, specific sulpho-transferases. The significance of this covalent modification is not obvious, although clearly it would introduce a negatively charged group into the molecule.

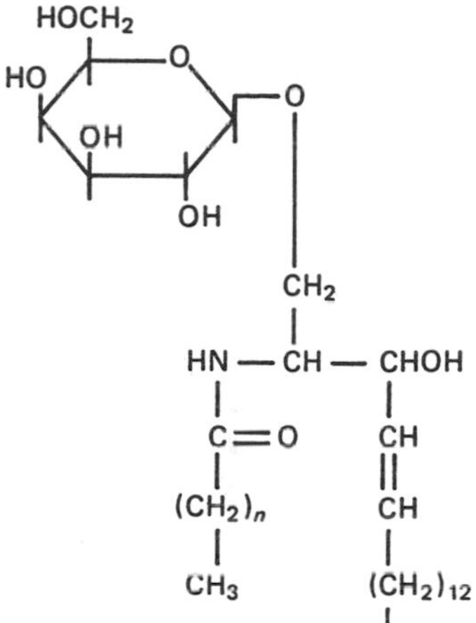

Galactocerebroside

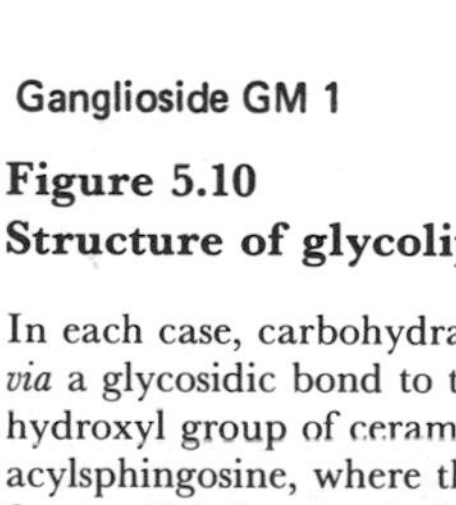

Galactose
|
N-Acetylgalactosamine
|
Galactose — Sialic
| acid
Glucose
|
Ceramide

Ganglioside GM 1

Figure 5.10
Structure of glycolipids

In each case, carbohydrate is linked *via* a glycosidic bond to the terminal hydroxyl group of ceramide (N-acylsphingosine, where the long-chain fatty acid is frequently lignoceric acid (C24:0). These glycolipids occur in particularly high concentrations in nervous tissue, where they are components of neuronal plasma membranes.

5.3   Intracellular protein sorting by the Golgi complex

5.3.1   Proteins with varying destinations leave the *trans*-Golgi network

Secretory proteins are synthesised on the rough ER and are somehow directed to their ultimate intracellular or extracellular destination. An overview of this process is presented in figure 5.11. A cell like a hepatocyte carries out a highly sophisticated sorting operation. Albumin is exported from the cell, receptors are incorporated into the plasma membrane, acid hydrolases are packaged into lysosomes, apolipoproteins become associated with lipoprotein particles. (In addition, resident proteins of the Golgi complex and the ER must be integrated into the appropriate organelle.) What are the mechanisms for achieving this complex sorting process? How are the proteins prevented from leaving their organelle? Which subcellular structure directs this intracellular traffic of secretory proteins?

Two assumptions are widely accepted by cell biologists: (1) secretory proteins carry a '*targeting signal*'; (2) this signal is recognised by a specific *receptor*, which directs the corresponding proteins to the appropriate subcellular site. (Lack of a targeting signal results in the export of that protein from the cell.) In only a few cases do we have convincing evidence for the interaction of a targeting signal of a secretory protein with a specific receptor protein. The soluble acid hydrolases of mammalian lysosomes have been one of the most widely studied systems; the targeting of these enzymes is considered in more detail in sections 6.4.2 and 6.4.3. We can state here though that the 215 kDa receptor for nascent acid hydrolases recognises that they bear a residue of mannose 6-phosphate, which is added post-translationally. Why only these enzymes are thus modified is not clear. Comparison of amino acid sequences derived from cloned cDNA does not reveal

Figure 5.11
Sorting of secretory proteins

Secretory proteins, synthesised on membrane-bound ribosomes, are translocated into the lumen of the endoplasmic reticulum (ER). Bulk flow will carry the proteins *via* the Golgi complex and the plasma membrane to the exterior of the cell, unless a targeting signal sequence on the protein is recognised by a receptor which traps that protein in a given subcellular compartment. The transport of proteins is represented by long thin arrows. The recognition/sorting events are represented by short thick arrows. It takes about an hour for a typical integral protein of the plasma membrane to travel from the *cis* Golgi to its final destination, and about 10 minutes from the rough ER to the *cis* Golgi.

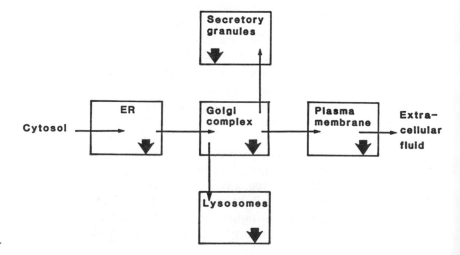

any primary structure common to the several acid hydrolases studied so far; probably the feature recognised is of a conformational nature (secondary or tertiary structure). The acidic interior of the lysosome causes the dissociation of the nascent acid hydrolase from its receptor, and thereby maintains the enzyme's subcellular localisation. In other cases, reverse transport may be prevented by selective proteolytic cleavage of a precursor form of the secretory protein within its organelle, so as to entrap the mature form of the protein.

Complex-type oligosaccharide chains are present on the glycoproteins of lysosomes, secretory granules, the plasma membrane, and the extracellular space. This observation implies that these proteins must have traversed the *trans* cisternae of the Golgi complex, where the appropriate glycosyl transferases are localised (table 5.1). From this point on, however, the route becomes more obscure. At what point does an extracellular protein become diverted from the pathway being followed by a lysosomal protein, or vice versa? Some information on this aspect has come from immuno-electron microscopic studies of Hep G2 cells (cultured hepatocytes), labelled with colloidal gold-conjugated antibodies to albumin (soluble extracellular protein) and mannose 6-phosphate receptor (lysosomal membrane protein). As shown in figure 5.12, the two proteins are initially present together in the small coated vesicles that are associated with the *trans*-Golgi network. At a

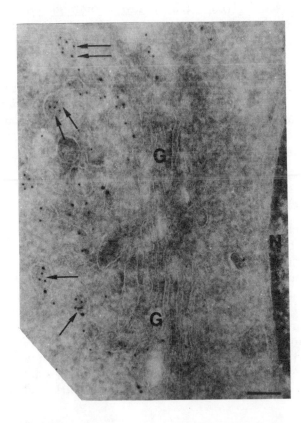

Figure 5.12
Sorting of proteins in the *trans*-Golgi network

The Golgi complex (G) is close to the nucleus (N). Antibodies to albumin were labelled with 6 nm particles of colloidal gold (small black dots), and antibodies to the mannose 6-phosphate receptor with 9 nm particles (large black dots). Initially, vesicles of the *trans*-Golgi network contain both proteins (single arrows). Subsequently, vesicles containing only albumin can be seen (double arrows). Bar = 0.1 μm. [Adapted from Geuze, H. J. *et al.* (1985). *J. Cell Biol.*, **101**, 2253–2262; reproduced by copyright permission of the Rockefeller University Press.]

Clathrin: Protein of molecular mass 180 kDa that forms trimers. These so-called triskelions associate to form a lattice or basket surrounding a variety of intracellular vesicles, particularly those involved in receptor-mediated endocytosis (see section 10.2).

more distal region, the proteins become segregated into separate populations of vesicles. These can be distinguished by the nature of their coat: the vesicles associated with the delivery of lysosomal proteins are enveloped with a scaffolding of **clathrin**, whereas the coated vesicles delivering extracellular and plasma membrane proteins are not.

Secretory granules release their contents by exocytosis in response to an appropriate extracellular stimulus. For example, *adrenocorticotrophic hormone* (ACTH) is packaged into vesicles of cultured AtT-20 cells by a process which involves self-aggregation of the hormone at slightly acidic pH. The resulting dense granules can be seen under the electron microscope (as in figure 5.5). Treatment of cells with the weak base *chloroquine* disrupts the packaging mechanism. Experiments with indicator dyes suggest that the interior of the *trans*-Golgi compartment is weakly acidic, around pH 6. Self-aggregation would account nicely for the observation that one can isolate a population of vesicles from AtT-20 cells that contain only ACTH. Further condensation of nascent secretory granules may even lead to crystallisation of the stored protein, as occurs in the case of insulin in pancreatic B cells.

A further complication exists in relation to *polarised epithelial cells,* which express different sets of proteins on their apical and basolateral surfaces. How is each plasma membrane protein guided to the appropriate surface? The answer to this question may come from studies of enveloped viruses (like vesicular stomatitis virus; section 4.2.3), which bud in a polarised manner from cultured canine kidney cells (figure 5.13). Viral glycoproteins follow a common route of transport up to and including the *trans*-Golgi compartment; sorting takes place somehow in the *trans*-Golgi network. Initial experiments with another protein, the polymeric immunoglobulin receptor, implicate a *C*-terminal signal for its incorporation into the aptical membrane.

In conclusion, secretory proteins are sorted into distinct populations of vesicles in the *trans*-Golgi network. Segregation of the different classes of protein probably involves a specific signal (sequence or conformation) for each. Lack of such a signal results in transport of that protein to the plasma membrane.

5.3.2   Vesicles mediate the intracellular traffic of secretory proteins

We have already seen how secretory proteins are packaged into distinct populations of vesicles (section 5.3.1). A similar selective capture of proteins occurs in the process of receptor-mediated endocytosis, considered in more detail in section 10.2. Transport vesicles ferry secretory proteins from the ER to the Golgi complex (section 4.2.3 and figure 5.7). For proteins that traverse the entire Golgi complex, what is the mechanism that transports them from one saccule to the next in a vectorial manner (*cis* → medial → *trans*)?

Cultured cells treated with metabolic poisons are not competent to transport nascent glycoproteins through the Golgi stack (as assessed by the appearance of complex-type oligosaccharide chains). Thus,

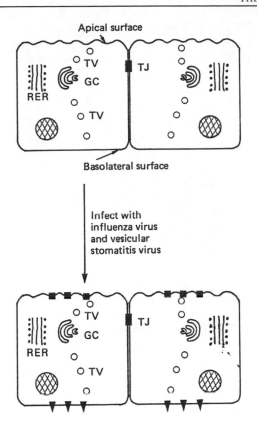

Figure 5.13
Viral glycoprotein targeting in polarised epithelial cells

In cultured canine kidney cells, the plasma membranes of the apical surface and of the basolateral surface are morphologically distinct. Membrane proteins are prevented from diffusing from one surface to the other by tight junctions (TJ) between adjacent cells. Virus-infected cells synthesise viral glycoproteins on the rough ER (RER) and export them to the plasma membrane *via* the Golgi complex (GC) and transport vesicles (TV). Immunofluorescence studies detect membrane glycoproteins of influenza virus (■) only at the apical surface, but those of vesicular stomatitis virus (▼) only at the basolateral surface.

direct transfer from one cisterna to another is not involved; indeed, direct connections between adjacent saccules have not been observed under the electron microscope. Instead, an ATP-dependent process is implicated. The nature of this process has been investigated by the elegant studies of **Rothman**, outlined in figure 5.14. The principle underlying these experiments was simple: in the fused hybrid cells, newly synthesised complex-type glycoproteins would only appear if their high-mannose precursors could transfer from their original Golgi complex (lacking galactosyl transferase) to that of the partner cell (which has the transferase). The results were unequivocal: complex-type glycoproteins *did* appear. The interpretation is that vesicles can shuttle nascent glycoproteins between one Golgi stack and another, by a process of budding and fusion. Indeed, swarms of small vesicles can often be seen under the electron microscope at the lateral edges of the Golgi complex (figure 5.2). Rothman is now further dissecting this transport process, which he calls 'hopping', in order to identify the components involved. How do the transport vesicles correctly recognise a neighbouring Golgi stack? How are resident Golgi proteins excluded during the budding process? We await the answers.

Clearly the same transport vesicles also mediate the vectorial transport of proteins, lipids and carbohydrates through the Golgi stack, in the direction *cis → trans*. Protein-laden vesicles leave the *trans*-Golgi

Figure 5.14
Vesicular transport between
Golgi complexes

Mutant cells (top left) lack galactosyl
transferase in their Golgi complex.
After infection with vesicular
stomatitis virus and incubation with
[^{14}C]-amino acid, the mutant cells
synthesise labelled viral G protein.
Wild-type cells (top right) possess
galactosyl transferase and are
incubated in the absence of virus or
radioactive amino acid. After fusion to
form a hybrid cell (centre), vesicles
transport G protein from one Golgi
complex to the other. The extent of
this transport can be assessed by
measuring the radioactivity associated
with galactosylated G protein.

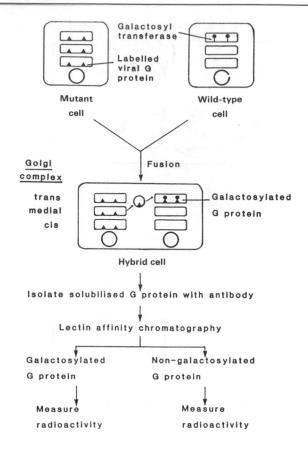

network to become lysosomes or secretory granules, or to fuse with the
plasma membrane in a process of exocytosis. There is also a flow of
vesicles in the opposite direction, with some of the vesicles fusing with
the lateral margins of the Golgi cisternae. As far as we know,
Golgi-derived vesicles do not travel in the direction *trans* → *cis*, nor do
they shuttle from the *cis* face of the complex to the ER. Thus, the Golgi
complex acts as a one-way valve, regulating the intracellular traffic of
vesicles.

5.4   Summary

Long ignored, the Golgi complex is now recognised as playing a key
role in the cell's synthesis of glycoproteins, the biogenesis of lysosomes
and other organelles, and the regulation of vesicular transport processes.
The stack of Golgi cisternae is structurally and functionally polarised
so as to facilitate these functions. The *cis* face receives nascent
glycoproteins from elements of the transitional ER. The glycoproteins
are acted on by glycosyl transferases and 'trimming' glycosidases
localised in specific sub-compartments of the Golgi complex. Other
post-translational modifications occurring in the organelle include
acylation, sulphation, and some of the stages of collagen biosynthesis.

Acid hydrolases acquire the mannose 6-phosphate marker which permits their segregation into primary lysosomes. This process occurs in the *trans*-Golgi network of tubules and vesicles, but we do not yet have a detailed picture of the complex sorting processes that direct secretory proteins to their ultimate intracellular or extracellular destination.

5.5 Study questions

1. Describe how the organisation of the Golgi complex is related to its role in glycoprotein processing. (section 5.2.1)
2. What experimental evidence suggests that the Golgi complex is organised into *cis,* medial, and *trans* sub-compartments? (figure 5.6)
3. Speculate on how galactosyl transferase comes to be localised in the *trans* cisternae of the Golgi complex. (section 5.3.1)
4. It has been proposed that nascent primary lysosomes bud off from the *cis* cisternae of the Golgi complex. What analysis of acid hydrolases could confirm or refute this proposal? (table 5.1)
5. Asialo-orosomucoid (ASOR, a glycoprotein lacking terminal sialic acid residues) is internalised by cultured hepatocytes, and re-emerges from the cells an hour later, this time with its oligosaccharide chains terminating in sialic acid. What does this result suggest about the route followed by the internalised ASOR? (table 5.1)
6. How might one confirm that the glycosyl transferases of the Golgi complex face the lumen of the cisterna, and not the cytosol? (sections 5.2.1 and 5.2.2).

5.6 Further reading

Goldfischer, S. (1982). *J. Histochem. Cytochem.,* **30**, 717–733.
  (Historical overview)
Rothman, J. E. (1985). *Sci. Amer.,* **253** (3), 74–89.
  (Specialisation of Golgi sub-compartments)
Griffiths, G. and Simons, K. (1986). *Science,* **234**, 438–443.
  (Sorting in the *trans*-Golgi network)
Pfeffer, S. R. and Rothman, J. E. (1987). *Ann. Rev. Biochem.,* **56**, 829–852.
  (Role of the Golgi complex in protein targeting)

6 THE LYSOSOME

6.1 Introduction

In the early 1950s **Christian de Duve** was working at Louvain University in Belgium, where he was interested in the biological role of the enzyme, *acid phosphatase* (figure 6.1). In order to study its subcellular distribution, he was applying the technique of differential centrifugation (section 2.2.1) to isotonic homogenates of rat liver. Curiously, he found that the activity of acid phosphatase in a freshly prepared homogenate was much less than that in a homogenate that had been left on the bench overnight or that had been frozen and thawed. These results led de Duve to formulate the concept of *structure-linked latency*: that inside the cell some enzymes were shielded from their substrate by an impermeable structure of some kind. What was the nature of this structure?

De Duve found that acid phosphatase activity was sedimented in the ultracentrifuge at centrifugal forces that brought down mitochondria, after nuclei had been removed and before microsomes came down (figure 6.2). Examination of the 'light-mitochondrial' pellet by electron microscopy revealed numerous profiles of mitochondria, interspersed with electron-opaque vacuoles having a single limiting membrane. That acid phosphatase was located in the latter structures was revealed by histochemical staining: enzymic release of inorganic phosphate from a phosphomonoester substrate was coupled to the deposit of an electron-dense product (lead phosphate). Furthermore, de Duve was able to purify the membrane-bound vesicles if the experimental rat was pre-treated with a detergent-like compound, Triton WR1339. This compound decreased the buoyant density of the vesicles such that they could be separated from mitochondria by isopycnic centrifugation in a sucrose density gradient (section 2.2.1). Further studies of purified vesicle preparations led to their characterisation in terms of:

(a) A single limiting membrane.
(b) A content of enzymes, all of which were hydrolases with an acidic pH optimum (typically around pH 5).
(c) Structure-linked latency, whereby the hydrolases were inaccessible

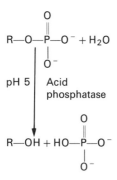

Figure 6.1
Reaction catalysed by acid phosphatase

The enzyme catalyses the hydrolysis of phosphate esters, with little specificity for the R group.

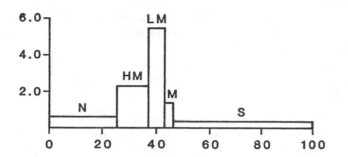

Relative specific activity

Proportion of total protein recovered (%)

Figure 6.2
Subcellular distribution of acid phosphatase

Enzyme activity was measured in fractions from an isotonic homogenate of rat liver that had been subjected to differential centrifugation. Fractions are labelled as follows: N, nuclear; HM, heavy-mitochondrial; LM, light-mitochondrial; M, microsomal; S, soluble supernatant (cytosol).

to their substrates unless the limiting membrane was first ruptured. This end could be achieved by physical insult (freeze-thawing or sonication) or by inclusion in the assay medium of detergents such as Triton X-100.

De Duve assigned the above characteristics to a subcellular organelle he called the *lysosome*. Its physiological function would be to degrade macromolecular material as a result of the concerted action of a battery of acid hydrolase enzymes (figure 6.3). Since potential substrates could not penetrate the lysosomal membrane directly, they would first have to be packaged inside a membrane-bound vesicle that subsequently fused with the lysosome. At the same time, the physical isolation of the hydrolytic enzymes would benefit the cell by protecting its macromolecular components from random degradation. Such considerations led to the early idea of the lysosome as a '*suicide bag*', whose disruption would bring about cell death. Such a concept is now considered naive, although it had a potent attraction at the time.

The development of the concept of the lysosome was enhanced by contemporaneous studies on the ultrastructure of kidney cells by **Strauss** and **Novikoff**. Under the electron microscope they could identify numerous '*dense bodies*', vesicles bounded by a single membrane and having an electron-opaque matrix. Histochemical staining for acid phosphatase located the enzyme inside these dense bodies. This experimental approach allowed cell biologists to differentiate lysosomes

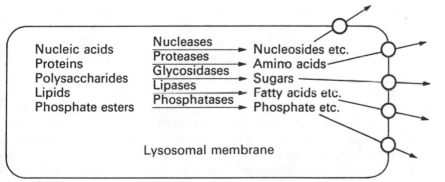

Nucleic acids
Proteins
Polysaccharides
Lipids
Phosphate esters

Nucleases
Proteases
Glycosidases
Lipases
Phosphatases

Nucleosides etc.
Amino acids
Sugars
Fatty acids etc.
Phosphate etc.

Lysosomal membrane

Figure 6.3
Degradation of macromolecules in the lysosome

Hydrolytic enzymes catalyse the breakdown of macromolecules to their component residues. Such low-molecular-mass solutes can pass across the lysosomal membrane *via* the appropriate transport proteins (represented by circles in the lysosomal membrane).

from other membrane-bound organelles, and also established the heterogeneous morphology of lysosomes, a feature closely related to their life-history (section 6.2)

So what are the central questions that biochemists now hope to answer concerning the structure and function of lysosomes? They include:

(a) How is a single class of enzyme, the acid hydrolases, segregated within a single class of subcellular organelle, the lysosome?
(b) What is the mechanism of biogenesis of lysosomes?
(c) How is the interior of the lysosome maintained at an acidic pH?
(d) How do lysosomes recognise the other membrane-bound components with which they may legitimately fuse?
(e) What physiological roles do lysosomes play?

Some of these questions are amenable to experimental approaches, others have been partly answered by the study of a group of genetic diseases that disrupt normal lysosomal function, the *lysosomal storage disorders*. First, however, we need to consider the morphology of the lysosome and its life-history.

6.2 The morphology and life-history of lysosomes

6.2.1 Lysosomes form a structurally heterogeneous class of organelles

The lysosome is normally identified in electron micrographs by histochemical staining for the marker enzyme, acid phosphatase (figure 6.4). Non-specific staining with heavy metals often reveals similar profiles of roughly spherical organelles bounded by a single membrane and having an electron-dense matrix. Modified versions of electron microscopic techniques can specifically label lysosomes by using anti-(acid hydrolase) antibody coupled to an electron-dense ligand such as colloidal silver or gold, or ferritin (figure 6.5). What kind of structure emerges from such observations? The over-riding impression is that of a diverse array of membrane-bound vesicles, ranging in diameter from 50 to 500 nm, and with heterogeneous contents. The matrix may be uniformly electron-dense, except for a pale rim just inside the limiting membrane. Alternatively, a largely electron-lucid interior may contain recognisable fragments of a mitochondrion or of endoplasmic reticulum (figure 6.6). In some cells the lysosomes may even take on a tubular morphology.

Where are lysosomes found? They are present in all eukaryotic organisms, it seems, from amoeba to Man. In mammals, they are present in the cells of all tissues and organs, with the exception of the red blood cell (and even that cell contains acid phosphatase as a relic of its development from a progenitor with organelles). There may be around 300 lysosomes in an average cell (figure 6.7), but many more times that number are present in those cells specialised for phagocytosis

200 nm

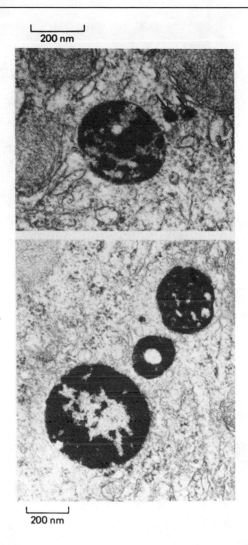

200 nm

Figure 6.4
Lysosomes of a mammalian
liver cell

In these thin-section electron
micrographs, lysosomes have been
stained by the Gomori procedure:
inorganic phosphate (released by the
action of lysosomal acid phosphatase
on β-glycerophosphate) is precipitated
as the electron-dense lead salt. [Taken
from B. Alberts *et al.* (1983) *Molecular
Biology of the Cell*; reproduced by
permission of Garland Publishing Inc.
Original photograph courtesy of
D. S. Friend.]

(section 6.5.2). Lysosomes are derived biosynthetically from elements of the *Golgi complex* (section 6.4.4), and the *trans* regions of that organelle do stain positively for acid phosphatase. Such observations led Novikoff to put forward the concept of the *Golgi-endoplasmic reticulum–lysosome* system (GERL) as a structurally and functionally interrelated set of organelles. We will see an example of their interaction when we consider the biosynthesis of lysosomal enzymes (section 6.4.2).

6.2.2   The lysosome family has a diverse life-history

The heterogeneous nature of the morphology of lysosomes is in turn a reflection of the diverse life-history of that organelle. What do we mean by this statement?

It is thought that new lysosomes are derived from the Golgi complex in a process that gives rise to small (typically 50 nm in diameter), electron-dense vesicles containing a full set of acid hydrolases and other

Figure 6.5
Lysosomes visualised by immuno-gold labelling

In this ultra-thin section electron micrograph of a macrophage cell, the lysosomal enzyme cathepsin D has been labelled by a specific antibody to which has been coupled colloidal gold particles (the small black dots). Most of the label is present over the profiles of secondary lysosomes. In one organelle (asterisk), the interior is unlabelled. Bar, 4 μm. [Taken from van Dongen *et al.* (1984). *Histochem. J.*, **16**, 941–954; reproduced by permission of the publisher, Chapman & Hall.]

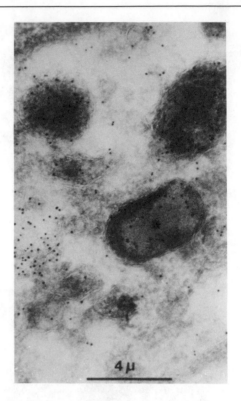

Figure 6.6
Autophagic vacuole of a mammalian liver cell

In this thin-section electron micrograph, a single large autophagic vacuole contains a diverse array of membranous and granular material undergoing digestion. [Taken from R. T. Dean (1977). *Lysosomes* (Studies in Biology no. 84), Edward Arnold Ltd, London. Original micrograph courtesy of K. Ogawa; reproduced by permission of the publisher.]

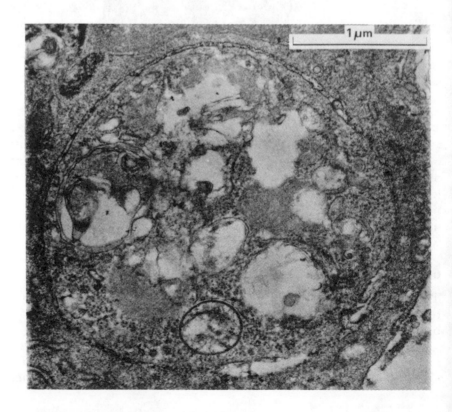

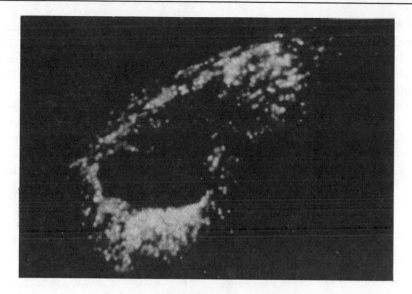

Figure 6.7
Visualisation of lysosomes by immunofluorescence microscopy

This fibroblast cell was treated with a rabbit antibody to lysosomal hexosaminidase and then with a goat anti-(rabbit IgG) antibody coupled to fluorescein. After ultra-violet illumination and microphotography, the fluorescence label is seen as white dots, which correspond to the lysosomes. These are dispersed throughout the cytoplasm, but are concentrated around the nucleus (the dark central mass). [Taken from van Dongen *et al.* (1984). *Histochem. J.*, **16**, 941–954; reproduced by permission of the publishers, Chapman & Hall.]

lysosomal proteins. Such 'virgin' lysosomes, which have not yet encountered another vacuole with which they might fuse, are generally referred to as *primary lysosomes* (figure 6.8). Once a primary lysosome has fused with another vacuole, and its enzymes have mingled with potential substractes, it is considered to be a *secondary lysosome* (figure 6.8). Now the other, substrate-laden, vacuole may be derived from several sources. In general terms, one can classify such vacuoles as being either **autophagic** or **endocytic**. Endocytic vacuoles may in turn be classed as *phagocytic* or *pinocytic,* depending on whether they contain primarily insoluble or soluble material, respectively (figure 6.8).

Autophagic: Containing intracellular material destined for degradation.
Endocytic: Containing material internalised from the extracellular space.

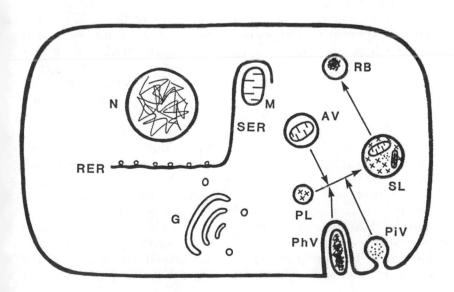

Figure 6.8
Life-history of the lysosome

In this diagrammatic representation, acid hydrolases (X) are synthesised on the rough endoplasmic reticulum (RER) from mRNA transcribed in the nucleus (N), and packaged by the Golgi complex (G) into a primary lysosome (PL). The formation of an autophagic vacuole (AV), containing a mitochondrion (M), of a phagocytic vacuole (PhV) and of a pinocytic vacuole (PiV), is shown. Fusion of these vesicles with a primary lysosome results in a secondary lysosome (SL), which may accumulate enough indigestible material to become a residual body (RB).

Thus, the life-history of the lysosome starts with its birth as a primary lysosome, and progresses as a secondary lysosome through multiple rounds of fusion with vacuoles containing macromolecular material destined for degradation. The heterogeneous nature of this material probably accounts in part for the diverse morphology of lysosomes. It is thought that eventually a secondary lysosome accumulates so much material that it cannot degrade, and/or loses too much of its enzymic complement through denaturation or internal proteolysis, that it becomes defunct. It is then referred to as a *residual body* (figure 6.8). Residual bodies may accumulate in cells or their contents may be extruded from the cell by exocytosis.

6.3 The composition of lysosomes

6.3.1 The lysosomal membrane has distinctive components, distinctive properties

The single membrane that surrounds the lysosomal matrix has a number of distinct properties that one has to account for in terms of its structure and composition. For example:

(a) It is permeable only to solutes of low molecular mass (less than 300 Da).
(b) It is responsible for the acidification of the lysosomal *matrix* (interior).
(c) It specifically recognises those cellular membranes with which it may legitimately fuse.
(d) It is resistant to attack by the hydrolase enzymes in the lysosomal matrix.

What do we know about the composition of the lysosomal membrane?

The major lipid components of the membrane are unexceptional, with phosphatidylcholine predominating (table 1.5). However, some of the minor components are unusual, and one of them, bis(mono-acylglycero)phosphate, appears to be unique to the lysosomal membrane. The function of this lipid is unknown. In view of the fact that components of the lysosomal membrane are, *via* the processes of endocytosis and exocytosis, continually mingling with those of the plasma membrane, it is surprising that the lipid composition of each is distinct (table 1.5). In particular, bis(monoacylglycero)phosphate is absent from the plasma membrane. Presumably there is some mechanism for the selective recapture of a distinct population of lipids when exocytosis is followed by endocytosis, or when small vesicles bud off from secondary lysosomes and return to the plasma membrane.

The membranes of a mixed population of lysosomes (primary and secondary) contain at least 16 different polypeptides that can be resolved by electrophoresis in SDS−polyacrylamide gel, with molecular masses ranging from 30 to 200 kDa. Many of the proteins are also glycoproteins. The function of most of the protein components of the lysosomal membrane is still something of a mystery, but one would predict the presence of certain acid hydrolases, of the lysosomal *proton pump*, of specific *permeases* (transport proteins), and of components that recognise other subcellular organelles or structures.

In a few cases, we have a detailed picture of the structure and function of a lysosomal membrane protein. For example, *acid β-glucosidase*, also known as glucocerebrosidase, catalyses the hydrolysis of the glycosphingolipid, glucosylceramide, at around pH 5 (figure 6.9). The enzyme is a glycoprotein which is only fully solubilised experimentally in the presence of detergent, a fact that suggests it is tightly associated with the lysosomal membrane. The biosynthesis of acid β-glucosidase was followed by pulse-chase studies (figure 2.10), which revealed that it is synthesised as a precursor of 62 kDa that is subsequently converted to a 5 kDa polypeptide by proteolytic processing and by addition of *N*-linked oligosaccharides to specific asparagine residues. The fine structure of the enzyme has been elucidated by cloning the human gene for acid β-glucosidase (section 6.4.1) and determining its base sequence and the derived amino acid sequence. A long open reading frame of 1545 bases codes for 515 amino acids, corresponding to a polypeptide of molecular weight 55 384. The first 19 amino acids (counting from the presumptive initiator methionine) at the *N*-terminal are highly hydrophobic, and thus represent the signal peptide that directs the nascent polypeptide across the membrane of the endoplasmic reticulum (section 4.2.1); there are five asparagine residues that reside in the appropriate sequence to act as sites for *N*-linked glycosylation (section 4.2.3); and there is a highly hydrophobic sequence near the *C*-terminal (...Ala.Val.Val.Val.Val.Leu... at residues 455–460 inclusive) that may be the enzyme's 'anchor' in the lysosomal membrane. In combination with classical protein chemistry studies using covalent inhibitors, these recombinant DNA studies indicate that the active site of the enzyme resides somewhere in the region of residues 428–461. However, it is not clear why the enzyme should be membrane-bound, rather than being freely soluble in the lysosomal matrix.

One question which perplexed cell biologists in the 1960s was the origin of the acidic nature of the intralysosomal environment; it has now largely been settled by the characterisation of a H^+-*ATPase* located in the lysosomal membrane. This enzyme is distinct from the functionally similar H^+-ATPases present in the mitochondrial inner membrane or in the plasma membrane of parietal cells of the stomach. We still know relatively little about the lysosomal enzyme, but it is clear that it can acidify the organelle's contents to a pH as low as 4.5. The proton pump is electrogenic, but the potential (positive inside) is largely neutralised by the subsequent uptake of Cl^- ions.

The products of the digestion of macromolecules by lysosomal hydrolases include amino acids, monosaccharides, fatty acids, cholesterol, nucleosides, and inorganic phosphate and sulphate, among others (figure 6.3). Such compounds, being relatively hydrophilic, would not be expected to diffuse freely across a lipid bilayer. Presumably some mechanism exists to facilitate their transport out of the lysosome. One

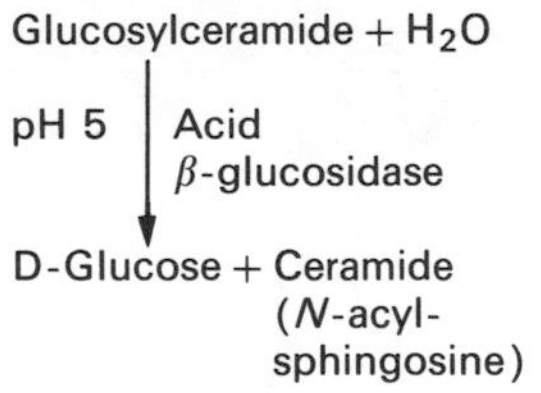

Glucosylceramide + H_2O

pH 5 | Acid β-glucosidase

D-Glucose + Ceramide (*N*-acyl-sphingosine)

Figure 6.9
Activity of lysosomal β-glucosidase

approach that has been adopted to study the kinetics of this process has exploited the fact that the lysosomal membrane is freely permeable to amino acid esters. Once inside the organelle, these compounds are hydrolysed by esterases to liberate the free amino acid, whose exit can be followed if the appropriate portion of the ester derivative was radioactively labelled. There are at least three permeases that facilitate the exit of amino acids and that have different substrate specificities. Although we know relatively little about the transport of solutes out of the lysosome, one can see the benefit of having an otherwise impermeant membrane: macromolecules must be completely degraded to their component parts, so that the latter can be made available to the biosynthetic machinery (and hence the economy) of the cell.

A number of proteins of the lysosomal membrane have been characterised with the aid of antibodies, whether monoclonal or polyclonal. At least three such proteins have been described in rat liver, with masses of 120, 100 and 80 kDa. In immunofluorescence microscopy studies, these antigens co-localised with the lysosomal marker enzyme, acid phosphatase, but were absent from other intracellular sites. The 120 kDa protein is heavily glycosylated, with at least 18 N-linked oligosaccharide chains rich in sialic acid. Perhaps this extensive glycosylation creates a 'glycocalyx' on the inner surface that serves to protect the components of the lysosomal membrane against attack by the acid hydrolases. Otherwise, the function of these protein antigens is largely unknown (although the 100 kDa glycoprotein may be the H^+-ATPase), as is the identity of those proteins that recognise other intracellular membranes with which a lysosome may legitimately fuse.

6.3.2 The lysosomal matrix is a store-house of acid hydrolases and other proteins

The major component of the lysosomal matrix is a single class of enzyme, the *acid hydrolase*. More than 50 such enzymes have been described, though not all are present in the lysosomes of every tissue, and a few are components of the organelle's membrane (section 6.3.1). The most abundant are present in concentrations estimated to be as high as 1 mM, when the concentration of the enzyme is most likely to be higher than that of its substrate! This has obvious advantages to the economy of the cell, since macromolecules destined for degradation will be broken down to completion very rapidly; also the product of one enzyme's action is likely to become the substrate of a nearby enzyme. Lysosomal enzymes themselves seem to be relatively resistant to attack by their own proteases, the cathepsins, and to the denaturing effects of the acidic environment. Table 6.1 presents a summary of lysosomal enzymes and their substrates.

Many acid hydrolases are 'exo' enzymes, which catalyse the hydrolysis of the terminal residue of a macromolecule. For example, the degradation of ganglioside GM2 requires the concerted action of several glycosidases, each removing a sugar residue in a step-wise manner from the non-reducing end (figure 6.10). The absence of one

Class of enzyme	Enzyme	Substrate(s)
Glycosidase	Acid maltase	Glycogen
	Hexosaminidase	Gangliosides, glycosamino-glycans, glycoproteins
Phosphatase	Acid phosphatase	Phosphomonoesters
Sulphatase	Arylsulphatase A	Sulphatide
Lipase	Acid lipase	Triglyceride
	Cholesterol esterase	Cholesterol esters
	Sphingomyelinase	Sphingomyelin
Protease	Cathepsin B, D, H, etc.	Proteins
Nuclease	Deoxyribonuclease	DNA
	Ribonuclease	RNA

Table 6.1
Some representative lysosomal enzymes and their substrates

enzyme in a pathway creates a block that results in the excessive accumulation of undegraded material, as seen in the lysosomal storage disorders (section 6.6). Acid hydrolases tend to be specific for the residue that they recognise and the bond which links it, but not for the macromolecule of which the residue is a part. Thus, hexosaminidase (table 6.1) is involved in the catabolism of the oligosaccharide moieties of several different classes of macromolecule, including glycosamino-glycans, gangliosides and glycoproteins. In contrast to such an 'exo' enzyme are the 'endo' enzymes, which catalyse the hydrolysis of internal bonds. Examples include glycosidases (acid maltase, hyaluro-nidase), proteases (various cathepsins) and nucleases (DNase, RNase).

Not all the proteins of the lysosomal matrix are enzymes. Some relatively small proteins (20 kDa or so) participate in the catalytic process by facilitating the interaction between a specific hydrolase and its substrate; the '*activator*' proteins involved in sphingolipid catabolism are some of the best understood. One other non-enzymic protein of the lysosomal matrix merits attention. It is a relatively small glycoprotein (32 kDa) that stabilises the association between two acid hydrolases: β-galactosidase and sialidase (neuraminidase). It seems that inside the lysosome, these enzymes aggregate together when in the presence of the stabilising protein. In its absence, as happens in one particular form of lysosomal storage disorder (section 6.6.1), both enzymes are rapidly denatured and degraded. The rationale behind this primitive form of multi-enzyme complex is unclear, although it might facilitate the breakdown of certain gangliosides, where β-galactosidase and sialidase act in sequence.

Other components are present in the lysosomal matrix. Secondary lysosomes contain the products of digestion of macromolecular material, at various stages of degradation. In some cases (the so-called 'residual bodies'), the matrix may be filled with an indigestible product called *lipofuscin*. In others ('multivesicular bodies') small vesicles are present, presumably bearing membrane-bound material for degradation. Thus, the heterogeneous nature of the lysosomal matrix is largely a reflection of the non-specific role of the organelle in cellular turnover.

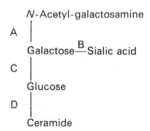

Figure 6.10
Stepwise degradation of ganglioside GM2

Shown here is the structure of ganglioside GM2. Degradation of this glycolipid is catalysed by lysosomal acid hydrolases. The enzymes, acting sequentially from the non-reducing end, are: A, *N*-acetyl-β-galactosaminidase (hexosaminidase);
B, neuraminidase (sialidase);
C, β-galactosidase; D, acid β-glucosidase.

6.4   The biogenesis of lysosomes

Any dividing cell must synthesise new primary lysosomes for transmission to its progeny, and many cells respond to a variety of stimuli by generating more lysosomes. Clearly, there is a controlled mechanism for the biosynthesis of acid hydrolases in the correct quantities and their packaging within a unique membrane to give a new primary lysosome. What is the nature of this biogenesis?

6.4.1   The genes for several lysosomal enzymes have been cloned

At the time of writing (1987) the genes for a number of acid hydrolases have been cloned: hexosaminidase A and B, acid β-glucosidase, and α-galactosidase A, among others. The approaches adopted by the biochemists involved varied considerably, but all had to circumvent a number of problems:

(a)   Any one lysosomal enzyme represents an extremely small proportion of the total cell protein (in the case of α-galactosidase A, estimated to be 0.002 per cent), and the level of its corresponding mRNA must be similarly small (perhaps only a few molecules per cell).

(b)   There is no tissue-specific distribution of lysosomes, so there is likely to be mRNA for lysosomal enzymes in all cells.

(c)   Likewise there are no physiological circumstances where one would expect dramatic changes in the levels of mRNA for acid hydrolases. (Points (b) and (c) together mean that one could not enrich a preparation of cDNA by excluding many of the non-lysosomal genes.)

(d)   Because of their extremely low concentrations, it is very difficult to obtain sufficient amounts of homogeneous lysosomal enzyme in order to carry out amino acid sequencing (with a view to constructing an oligonucleotide probe).

The usual approach has been to construct a cDNA library, either in plasmids (pBR322) or in phage (lambda gt11), and to screen for the presence of positive recombinant clones, by using either a monospecific antibody (in expression vector systems) or an oligonucleotide probe (where amino acid sequence data were available). For example, **Beutler** cloned the cDNA corresponding to acid β-glucosidase in an expression vector system (see section 2.3 for details). In the initial screen of such a cDNA library, only one clone in a million was positive, an indication of just how rare the corresponding mRNA species must have been. Proof that Beutler was indeed dealing with the gene for acid β-glucosidase came from sequencing studies: an open reading frame contained a codon sequence that matched some early protein sequencing data.

The genes for many more lysosomal proteins will be cloned in the near future. However, we currently have very few data on the fine structure of the gene of any acid hydrolase as it exists in the genomic DNA. There is some evidence that the genes for the subunits of hexosaminidase contain introns, but we know nothing of the DNA sequences that might control the transcription of the genes for this or any other acid hydrolase.

6.4.2 Lysosomal enzymes are synthesised as larger precursors

After transcription of the gene, the mRNA for each lysosomal enzyme binds a ribosome at the initiation site. Translation then proceeds in the manner as described in chapter 4 (section 4.2.2) for a secreted protein. The first 10–20 residues form a highly hydrophobic stretch (the signal peptide). There does not appear to be any homology between the few signal peptides that have been sequenced for lysosomal enzymes; presumably the hydrophobic nature of the N-terminal provides sufficient specificity. (To give just one example, that for acid β- glucosidase is Met.Ala.Gly.Ser.Leu.Thr.Gly.Leu.Leu.Leu.Leu.Gln. Ala.Val.Ser.Trp.Ala.Ser.Gly.) After cleavage of the signal peptide in the lumen of the ER, co-translational transport of the remainder of the polypeptide occurs across the membrane. For many lysosomal enzymes, the completed protein exists as a catalytically inactive pro-enzyme; presumably the cell would be damaged by acid hydrolase activity expressed in any compartment prior to the primary lysosome.

While in the lumen of the ER, the nascent enzyme undergoes glycosylation (section 4.2.3), with oligosaccharide chains added *en bloc* to asparagine residues from a dolichol-based donor. Although the specific sites of glycosylation have not been identified for any one enzyme, the gene sequencing data predict several candidate targets in the case of acid β-glucosidase: there are five sequences of asparagine.X (any amino acid).serine (or threonine) in the polypeptide. *O*-glycosylation of acid hydrolases appears to be a much less common event.

The glycosylated pro-enzyme is then transported in transport vesicles to the Golgi complex, where a number of covalent modifications take place (section 5.2.1). Some of the 'high-mannose' oligosaccharide chains undergo trimming, followed by terminal glycosylation to give 'complex' type chains. A small proportion of the mannose residues undergo a unique phosphorylation that provides the enzyme with a 'tag' which labels it as destined for a primary lysosome. The process involves two enzyme-catalysed steps, as depicted in figure 6.11. The result is a lysosomal pro-enzyme bearing one or more residues of mannose 6-phosphate, which acts as a marker for a receptor protein involved in the targeting of newly synthesised acid hydrolases. What is not clear is the nature of the structural marker on the lysosomal enzyme that allows it to be so modified, but not other proteins in transit through the Golgi complex.

Figure 6.11
Biosynthesis of the mannose 6-phosphate marker of lysosomal enzymes

The symbols used are ⬛L -, glycosylated lysosomal pro-enzyme; P, phosphate.

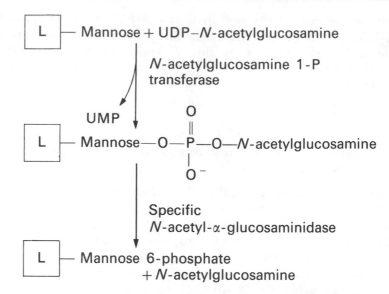

6.4.3 A pH-sensitive receptor delivers acid hydrolases to nascent primary lysosomes

The mannose 6-phosphate receptor of fibroblast cells is a large protein (215 kDa on SDS–polyacrylamide gels), that can be detected by immunocytochemistry in elements of the Golgi complex and (to a lesser extent) in the plasma membrane, but surprisingly not in lysosomes. Presumably the receptor cycles between the Golgi complex and nascent primary lysosomes, but spends most of its time in the former. The 215 kDa receptor binds tightly to lysosomal pro-enzymes at neutral pH in the absence of divalent metal cations. This affinity drops sharply below pH 6.0. Formation of a nascent primary lysosome is associated with its acidification (by the membrane-bound proton pump; section 6.3.1), and the acidic environment so created assists the mannose 6-phosphate receptor to unload its bound hydrolase, thereby allowing the receptor to recycle back to the Golgi complex, for another round of transport. This delivery process is disrupted by **lysosomotropic amines**.

Not all cells have the 215 kDa receptor, however, and not all lysosomal enzymes contain mannose 6-phosphate. There is evidence for an independent delivery pathway that involves a receptor (a trimer of 46 kDa subunits) with an absolute requirement for divalent metal cations, and there may also be a pathway independent of mannose 6-phosphate. Future work should establish the nature of the structural signals on those acid hydrolases that bind to the latter receptors. Membrane-bound enzymes such as acid β-glucosidase may follow the non-mannose 6-phosphate pathway.

Lysosomotropic amines: These are bases such as ammonia and chloroquine. Such compounds accumulate in nascent primary lysosomes, where they prevent adequate acidification. Thus, acid hydrolases *en route* to primary lysosomes are lost to other subcellular compartments and into the extracellular space. This trick — of incubating cultured cells with ammonia — is widely used by cell biologists to disrupt processes which are thought to involve an essential acidification step.

6.4.4 Primary lysosomes arise from a *trans*-Golgi network

The foregoing discussion accounts for the delivery of newly synthesised acid hydrolases to nascent primary lysosomes, but the question remains: whereabouts in the cell do these events occur, and at what stage do the lysosomal pro-enzymes become activated? The current consensus among cell biologists implicates the trans-*Golgi network* (TGN; figures 5.2 and 5.12). This is assumed to be the site where acidification of nascent primary lysosomes occurs so as to facilitate unloading of receptor-bound hydrolases. What is not clear is how only acid hydrolases come to be sequestered inside a vesicle with a unique membrane. There must be a highly sophisticated sorting operation taking place inside the TGN.

Most (though not all) of the pro-enzyme forms of the acid hydrolases are catalytically inactive, and require activation by partial proteolysis, in much the same way as that required by pro-insulin, trypsinogen or many other extracellular proteins. The enzyme involved in the activation of pro-hexosaminidase is probably also lysosomal, and is a thiol protease. We do not yet know whether this protease is able to activate all lysosomal pro-enzymes, or what the structural features are that it recognises.

6.5 The physiological functions of lysosomes

So far, we have considered in some detail the structure and biogenesis of lysosomes, but we have said little about the physiological roles of the organelle, except in the most general terms (section 6.1). Let us now redress the balance before we go on to consider the problems that arise when the normal functioning of lysosomes is disrupted (section 6.6).

6.5.1 Degradation of extracellular material aids the cell's economy

The cell is continually sampling its extracellular environment, by internalising it and subjecting its components to degradation in secondary lysosomes. In its simplest form, this process involves *pinocytosis* of the external fluid into smooth endocytic vesicles. Fusion with a lysosome permits mingling of the extracellular contents with the acid hydrolases under conditions that lead to degradation of the internalised macromolecular material. The products of this digestion process are released into the cytoplasm to contribute to the biosynthetic economy of the cell. It is in this manner that simple organisms such as amoebae use *digestive vacuoles* in order to provide sources of metabolic energy and some of the building blocks for their biosynthetic requirements.

A more specific pathway is followed by some components that have corresponding receptors in the plasma membrane. Uptake by *receptor-mediated endocytosis* leads to internalisation of the ligand inside clathrin-coated vesicles (section 10.2). The receptor is recycled to the plasma membrane, whereas the ligand undergoes degradation in the lysosome. In some cases, the digestion products are particularly important. For example, in humans cholesterol is delivered to peripheral cells in the form of its ester present in *low-density lipoproteins* (section 10.3). Receptor-mediated uptake of the lipoprotein is followed by the lysosomal hydrolysis of the cholesterol ester, thereby releasing the cholesterol (and concomitantly an essential, polyunsaturated fatty acid) for use by the cell.

6.5.2 Some cells are specialised for lysosomal digestion

Although all eukaryotic cells contain lysosomes (with the exception of mammalian erythrocytes), some cells with large numbers of lysosomes have a specialised role in the degradation of extracellular material. The classic examples are the professional phagocytic cells of the mammalian reticulo-endothelial system, present in the blood, lymph, liver, spleen and elsewhere. Macrophages are particularly adept at ingesting micro-organisms that have become **opsonised** (figure 6.12).

Opsonised: Coated with specific antibodies which recognise surface antigenic determinants.

Figure 6.12
Involvement of lysosomes in phagocytosis

A dividing bacterium is about to be engulfed by a macrophage cell, in this thin-section electron micrograph. The resulting phagocytic vacuole will fuse with lysosomes, the electron-dense vesicles at the bottom of the picture. [Taken from B. Alberts *et al.* (1983). *Molecular Biology of the Cell*; reproduced by permission of Garland Publishing Inc. Original micrograph courtesy of D. Bainton.]

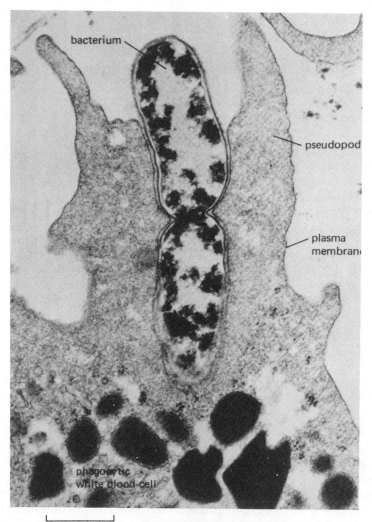

1 μm

Fusion of the phagocytic vacuole with lysosomes leads to the hydrolytic destruction of the internalised micro-organism. It seems that this process may not go to completion. Fragments of degraded micro-organism are presented by the macrophage, in a little understood fashion, to specific T-lymphocytes, which are then primed to recognise the intact micro-organism and to destroy it by a number of mechanisms. Thus, lysosomes may play an important role in the immune response.

6.5.3 Lysosomes are involved in the turnover of the cell's own constituents

In the preceding two sections we have seen how lysosomes are active in the degradation of material brought into the cell by endocytosis. Another normal aspect of the function of lysosomes is the breakdown of the cell's own constituents, whether soluble or insoluble. This process is all part of the normal turnover of the cell, whereby defunct components are continually replaced by newly synthesised ones. Since macromolecular material cannot penetrate directly into lysosomes, it must first be packaged inside a membrane-bound vesicle which subsequently fuses with a lysosome. This process is generally referred to as *autophagy*. The source of the enveloping membrane is not clear (though it may be the endoplasmic reticulum), neither is the mechanism whereby specific components are ear-marked for destruction. However, even organelles can be seen to undergo degradation in large secondary lysosomes (figure 6.6). At the molecular level, a proportion of the glycogen in the cytosol of liver and muscle cells undergoes hydrolysis to glucose *via* a lysosomal route, catalysed by acid maltase (α-glucosidase). Indeed, the components of the lysosome itself probably undergo slow self-degradation, catalysed by the surrounding acid hydrolases present in close proximity and at high concentration.

We have considered here, in this and in the preceding section, only a small selection of the roles played by lysosomes in the degradation or turnover of extracellular components (heterophagy) and intracellular components (autophagy). There are countless other examples. Particularly noteworthy is the involvement of lysosomes in the remodelling of tissues associated with embryonic development (in many species); with metamorphosis (in the frog, as the tadpole's tail regresses); with bone formation (where old bone is degraded by osteoclast cells); in kidney cells, where proteins that have passed from the blood plasma into the glomerular filtrate are taken up by endocytosis and degraded in secondary lysosomes); and in mechanisms of action of hormones (release of thyroid hormone by lysosomal digestion of thyroglobulin, or down-regulation of insulin receptors by lysosomal degradation, for example).

6.6   The pathology of lysosomes

We have already seen in several preceding sections how the study of defects in lysosomal metabolism has provided information about the normal functioning of this organelle. In many cases, the defect is a genetic deficiency of one particular component of the lysosome, but there are also examples of environmental insult that lead to improper lysosomal function. We will now consider these pathological aspects since they shed light on the normal structure and function of lysosomes.

6.6.1 Lysosomal storage disorders are progressive, generalised, genetic diseases

Pompe disease: Inherited deficiency of the lysosomal enzyme, acid maltase (α-glucosidase).

On the basis of his experience with **Pompe disease, Hers** put forward in 1965 the concept of *lysosomal storage disorders*. The distinguishing features of this class of inherited disease are:

(a) They are generalised disorders affecting all organs (since all tissues have lysosomes).

(b) They are progressive disorders, with the pathological symptoms becoming increasingly severe with age.

(c) They are associated with the accumulation of undegraded material in secondary lysosomes (hence their name).

(d) The undegraded material is the substrate for a specific lysosomal enzyme that is deficient in patients with a given disorder.

(e) The stored material is likely to be heterogeneous, since many acid hydrolases do not act specifically on a single class of macromolecule.

(f) The disorders are genetic conditions, inherited (usually) in an autosomal recessive manner.

Subsequent studies have amply confirmed Hers' original proposals, and there are now more than 30 lysosomal storage disorders that have been characterised in humans and other animals. Let us consider one such disorder, *Tay–Sachs disease,* and see how it illuminates the molecular pathology of these biochemically related conditions.

Tay–Sachs disease is one of a small group of lysosomal storage disorders that result in the excessive accumulation of a glycolipid, ganglioside GM2 (figure 6.10). The pathological consequences include secondary lysosomes engorged with undegraded membranous material (figure 6.13), severe mental and motor retardation, and death in early childhood. The enzyme that catalyses the hydrolysis of the terminal *N*-acetylgalactosamine residue of ganglioside GM2, *hexosaminidase A,* is deficient in tissues from Tay–Sachs patients. On the basis of this finding, biochemists and their clinical colleagues have devised reliable methods for the detection of individuals homozygous and heterozygous for Tay–Sachs disease. Such methods have been successfully applied to family genetic counselling, mass screening programmes, and pre-natal diagnosis of at-risk pregnancies.

The explanation proposed for the enzyme defect was that human hexosaminidase A was a dimer with two different subunits, α and β; Tay–Sachs disease would involve a genetic mutation in the α subunit responsible for the hydrolysis of ganglioside GM2. Somatic cell hybridisation studies confirmed that the genes for the two subunits

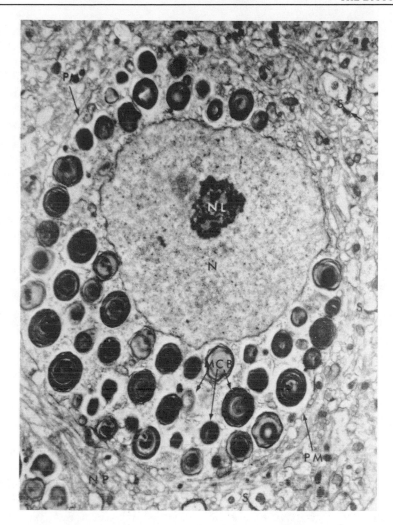

Figure 6.13
Secondary lysosomes in Tay–Sachs disease

This thin-section electron micrograph shows a neurone from the brain of a patient with Tay–Sachs disease. The cytoplasm is filled with secondary lysosomes (MCB) engorged with whorls of undegraded lipid material. Other subcellular structures are indicated as follows: N, nucleus; NL, nucleolus; PM, plasma membrane; S, synapse. × 9800. [Taken from R. D. Terry and M. Weiss (1963). *J. Neuropath. Exp. Neurol.*, **22**, 18–55; reproduced by permission of the journal's Association.]

were independent; that for the α subunit was located on human chromosome 15, whereas that for the β subunit was on chromosome 5. Pulse-chase experiments (figure 2.10) with monospecific antiserum to α or β subunit, detected mature β subunit in fibroblasts from Tay–Sachs patients, but no α subunit. The genetic lesion which causes at least some forms of Tay–Sachs disease results in lack of the 1.9 kb mRNA for the α subunit, as determined by Northern blotting with a cloned probe. Thus, all the biochemical and genetic information paints a coherent picture of human hexosaminidase in health and disease, and such work established a pattern that has been followed in studies of many other lysosomal storage disorders.

In principle, a patient suffering from a lysosomal storage disorder should respond well to **enzyme replacement therapy.** Sadly, such promise has not been borne out in the few clinical trials that have been attempted. A more radical approach to treatment now lies at the frontiers of current research: *gene replacement therapy.* In principle, it

Enzyme replacement theory:
Proteins infused into the bloodstream are taken up by cells by receptor-mediated endocytosis (section 10.2), before being sequestered in secondary lysosomes. Thus, if sufficient amounts of the appropriate pure human enzyme could be given to a patient, it should find its way to the precise site of the biochemical lesion and correct it.

Promoter: Nucleotide sequence, usually on the 5′ side of a gene, which activates the transcription of the gene.

should be possible to prepare large amounts of the human gene, probably in the form of cDNA, for insertion into deficient cells. One would have to genetically engineer the added gene so that it lay 'downstream' from a strong **promoter**, since otherwise the random incorporation of the exogenous DNA into the host genome would almost certainly result in a low level of expression. Also, one would like simultaneously to enhance the viability of the transformed cells, so that they were not displaced by overgrowth by the deficient cells. Preliminary reports have described the transformation of cells genetically deficient for acid β-glucosidase (the enzyme missing in *Gaucher disease*) by incorporation of the human gene in cDNA form. We await developments with interest.

In the large majority of the lysosomal storage disorders that have been studied to date, the genetic mutation has resulted in the deficiency of one particular acid hydrolase. However, this is not the only mechanism that can give rise to such pathology. Other lesions include:

(a) Lack of the 'activator' protein (section 6.3.2) that mediates the interaction between a lysosomal enzyme and its substrate *in vivo*.

(b) Lack of a 'stabiliser' protein (section 6.3.2) that facilitates the association of some acid hydrolases.

(c) Deficiency of a transport protein in the lysosomal membrane (section 6.3.1) that is required for the export of a particular solute from the organelle.

(d) In *I-cell disease* (mucolipidosis II), there is a multiple deficiency of many acid hydrolases in fibroblasts and other cells, and corresponding massive increases in the concentration of these enzymes in extracellular fluids. However, the genetic lesion does not affect the lysosomal proteins themselves, but rather it leads to a deficiency of the specific transferase in the Golgi complex that 'tags' nascent acid hydrolases with the mannose 6-phosphate marker (figure 6.11). The enzymes are thus not packaged correctly into primary lysosomes, and consequently are lost from the cell.

6.6.2 Some causes of lysosomal pathology are environmental in nature

What we have considered in the preceding section are examples of lysosomal pathology that result invariably from a deficiency of a specific component of that organelle. However, there are numerous other conditions where the normal physiological role of the lysosome is compromised in a different manner. In particular, a number of environmental factors have been implicated in disease states that lead to inappropriate expression of acid hydrolase activity.

One such example is *asbestosis*. In this condition small particles of asbestos inhaled into the lungs become internalised by macrophage cells there and pass into secondary lysosomes. Here, in some poorly defined way, they destabilise the lysosomal membrane, thereby causing a massive release of acid hydrolases. Over prolonged periods of time, the enzymes cause destruction of the cells lining the alveoli where gaseous exchange takes place, and the victim develops severe breathing problems. There is no effective treatment, but governments are now aware of the safety hazards posed by asbestos fibres, and have introduced legislation to control the use and removal of the material.

A similar kind of pathology is thought to underlie *rheumatoid arthritis* and *osteo-arthritis*, two increasingly common problems characterised by the inappropriate degradation of connective tissue in joints and bone, respectively. Although the underlying causes of these degenerative conditions are largely unknown, it has been appreciated for many years that patients often respond favourably to treatment with steroid drugs, such as hydrocortisone. These compounds are known to stabilise the membrane of lysosomes, and their clinical efficacy may result in part from this effect.

Lysosomal pathology may also be induced experimentally, both *in vitro* and *in vivo*. Mice have been infused with *conduritol epoxide,* a potent and specific inhibitor of acid β-glucosidase, with which it forms a covalent derivative at the enzyme's active site. The result is an animal model for Gaucher disease, a genetic deficiency of the enzyme. Inactivation of the hydrolase causes the intralysosomal accumulation of undegraded glucosylceramide (figure 6.9) and the development of many of the pathological symptoms seen in the corresponding human genetic condition. Such experimentally induced lysosomal storage disorders, and other genetic counterparts in animals, will probably be exploited for the purposes of testing modified versions of enzyme replacement therapy, as well as (ultimately) gene replacement therapy.

6.7 Summary

We have seen how our ideas of the structure and function of lysosomes have progressed from the early concept of the 'suicide bag' to the present model of a dynamic organelle with a key role in the turnover of the cell's macromolecules and in related activities. Another early concept, that of the Golgi–endoplasmic reticulum–lysosome complex, has been revived recently in a modified form in order to emphasise the close biosynthetic and functional links between these subcellular compartments. The acid hydrolases have been intensively studied in order to analyse their degradative functions and their involvement in the lysosomal storage disorders. Other components of lysosomes, such as the membrane permeases and the surface proteins that recognise other subcellular structures, remain by comparison largely uncharacterised. The source of new primary lysosomes has been identified, by techniques that combine the specificity of antibodies with the high resolution of electron microscopy, to a complex, ill-defined membranous network at the *trans* face of the Golgi complex. The signal for incorporation of newly synthesised acid hydrolases into a nascent primary lysosome is in many cases a residue of mannose 6-phosphate.

The genes for several acid hydrolases have been cloned in the form of cDNA, despite the difficulties involved. Future work should shed new light on the interrelationships between the structure and function of lysosomal proteins, and on the precise nature of the genetic lesions in the lysosomal storage disorders.

6.8   Study questions

1. Explain how you might prepare experimentally a highly purified preparation of lysosomes from an isotonic tissue homogenate. (section 6.1).
2. How would you identify a lysosome in a thin section of tissue prepared for transmission electron microscopy? (section 6.2.1).
3. Explain the term 'structure-linked latency' as applied to the activity of lysosomal enzymes. (section 6.1).
4. What are the structural and functional relationships between lysosomes, the Golgi complex, and the endoplasmic reticulum? (sections 6.2.1 and 6.4.4).
5. Assess the contribution made by the following techniques to our understanding of the biosynthesis of lysosomal enzymes: pulse-chase (radioactive labelling) experiments; cDNA cloning; immuno-electron microscopy.
6. Explain the molecular pathology that gives rise to a named lysosomal storage disorder. (section 6.6.1)
7. The concentration of soluble cathepsin D (molecular mass of 100 kDa) in a lysosome (diameter 200 nm) has been estimated at 1 mM. How many molecules of the enzyme does a single such lysosome contain? What does this enzyme concentration represent in mg of protein per ml? (Avogadro's number $= 6 \times 10^{23}$)

6.9   Further reading

Books

Dean, R. T. (1977). *Lysosomes*, Edward Arnold, London.
Pitt, D. (1975). *Lysosomes and Cell Function*, Longman, New York.

Reviews and specific articles

de Duve, C. (1983). *Eur. J. Biochem.*, **137**, 391–397.
    (Historical account of the structure and function of lysosomes, by their discoverer)
von Figura, K. and Hasilik, A. (1986). *Ann. Rev. Biochem.*, **55**, 167–193.
    (The biosynthesis of lysosomal proteins)
Lloyd, J. B. and Foster, S. (1986). *Trends Biochem. Sci.*, **11**, 365–368.
    (The lysosomal membrane and its transport proteins)
Griffiths, G. and Simons, K. (1986). *Science*, **234**, 438–443.
    (Revival of the GERL concept)

Sorge, J., West, C., Westwood, B. and Beutler, E. (1985). *Proc. Nat. Acad. Sci. USA*, **82**, 7289–7293.
   (Cloning of the human gene for acid β-glucosidase)
von Figura, K. and Hasilik, A. (1984). *Trends Biochem. Sci.*, **9**, 29–31.
   (Lysosomal enzymes and storage disorders)

7 THE PEROXISOME

7.1 Introduction

For many years after its biochemical characterisation by **de Duve** in the 1960s (section 1.2.2), the peroxisome was little more than a curiosity. Its sedimentation in the ultracentrifuge could be followed by the presence of the marker enzyme, *catalase*. Under the appropriate conditions of differential centrifugation and density-gradient centrifugation, the peroxisomes of rat liver could be separated from lysosomes, mitochondria, and microsomes. Biochemical analysis revealed a content of several oxidative enzymes; the feature common to all these enzymes was that they generated hydrogen peroxide (H_2O_2) as a product of their reaction (figure 7.1). Thus, it was considered that the peroxisome had developed in the course of evolution in order to protect the cell from the potentially harmful effects of these particular oxidative enzymes.

Furthermore, the single limiting membrane of the peroxisome, as seen under the electron microscope (figure 7.2), was thought to be 'leaky'. After isolating the organelles from rat liver, de Duve noted

Figure 7.1
Oxidative reactions in the peroxisome

The reactions are catalysed by the appropriate FAD-linked enzyme: 1, urate oxidase; 2, xanthine oxidase; 3, L- and D-amino acid oxidases. The potent oxidising agent, hydrogen peroxide, which all three reactions generate, is rendered harmless by the highly active haem-containing catalase (enzyme 4).

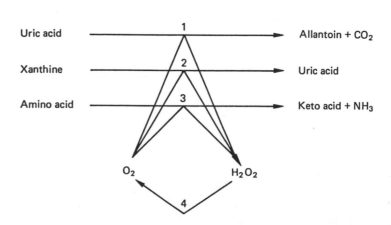

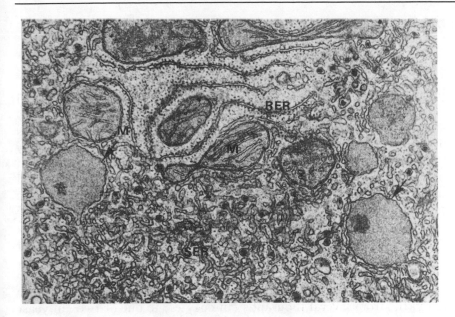

Figure 7.2
Peroxisomes in a liver cell

In this thin-section electron
micrograph, two large peroxisomes
(arrowed) are readily identified by
their crystalline-like core. Two smaller
peroxisomes lie close above that on
the right. Note the proximity to
mitochondria (M), and rough and
smooth endoplasmic reticulum (RER
and SER, respectively). [Taken from
Fawcett, D. W. (1981). *The Cell*,
2nd edn, W. B. Saunders Co.,
Philadelphia. Original micrograph
courtesy of R. Bolender.]

that catalase did not display the property of structure-linked latency,
which is a characteristic feature of the enzymes of lysosomes and other
membrane-bound vesicles. Hence he considered that the substrates for
the oxidative enzymes could diffuse freely into the peroxisome, and the
products could pass out likewise. The membrane itself was thought to
be derived from the smooth endoplasmic reticulum, where the peroxisomal
enzymes somehow segregated together within a vesicle. Support for
this hypothesis came from the electron microscopic studies of **Novikoff**,
who described the apparent continuity of the smooth ER with
an extention of the peroxisomal membrane.

These early ideas are now considered to be naive. The lack of
structure-linked latency was probably a result of damage to the
peroxisomal membrane in the course of the isolation procedure. The
matrix of the organelle is slightly acidic (about pH 6), a feature that
is inconsistent with a 'leaky' membrane (since otherwise protons would
diffuse out and the pH gradient would collapse). New peroxisomes do
not form *de novo* from the smooth ER (discussed further in section 7.3):
the lipid and protein composition of the membranes concerned are
distinctly different. As well as the oxidative enzymes (figure 7.1), the
peroxisome is now known to contain numerous other enzymes of both
a biosynthetic and a degradative nature. Lack of the organelle, as
occurs in certain inherited disorders in Man, is incompatible with life.
All in all, the peroxisome is more than just a curiosity; it makes a
significant contribution to the metabolism of most cells. The biogenesis
of the organelle is currently a topic of great interest among cell biologists,
since it probably involves in part some kind of fission process. First, let
us consider further the structure and functions of peroxisomes.

7.2   Structure and functions of peroxisomes

7.2.1   Peroxisomes are catalase-containing vesicles

Peroxisomes appear in electron micrographs as roughly spherical vesicles bounded by a single membrane and with a slightly electron-dense, granular matrix (figure 7.3). In mammalian organs such as liver and kidney, where they are relatively plentiful (table 1.1), they have diameters of $0.5-1.5$ μm. In other animal tissues, they are much smaller (about 0.1 μm diameter), and so are sometimes called 'microperoxisomes'. In plants, certain peroxisomes are referred to as *glyoxysomes*, since they contain the enzymes of the glyoxylate pathway (section 7.2.3). (In the older literature the term *microbody* is frequently applied to all those organelles which contain catalase on the basis of histochemical staining and electron microscopy. All these names refer to the same organelle.)

Peroxisomes are present in all eukaryotic cells. The number per cell varies from several thousands (in oocytes) to one or two (in yeast grown on sucrose). Electron microscopy often reveals clusters of peroxisomes adjacent to the endoplasmic reticulum. Despite earlier claims, no direct continuity has been seen between a peroxisome and the ER. Serial sectioning of mouse liver has, however, revealed continuities between the membranes of adjacent peroxisomes, thereby giving rise to extended (4 μm), convoluted tubular structures (the so-called 'peroxisomal reticulum'). Such observations have important implications for theories of the biogenesis of peroxisomes (section 7.3).

Figure 7.3
Morphology of peroxisomes

The single membrane and granular matrix are clearly visible in this high-magnification electron micrograph of a cluster of peroxisomes in a thin section of rat liver. The plane of section passes through the para-crystalline core of urate oxidase in the two lower organelles. These peroxisomes also have protuberances of their limiting membrane; that at lower left has in addition a tubular extension of its membrane. [Taken from Fawcett, D. W. (1981). *The Cell*, 2nd edn, W. B. Saunders Co., Philadelphia. Original micrograph courtesy of D. S. Friend.]

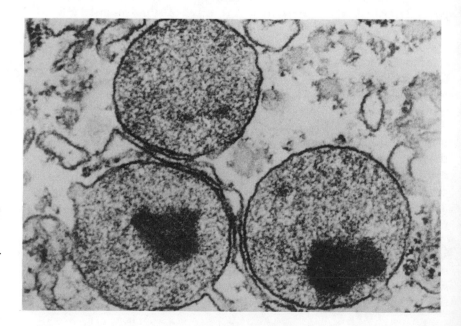

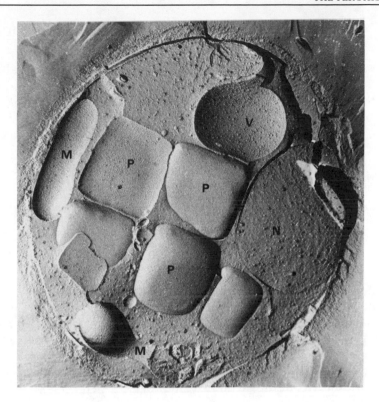

**Figure 7.4
Induction of peroxisomes in
yeast**

When the yeast *Hansenula polymorpha* is
grown in a medium containing
methanol as the sole carbon source,
the peroxisomes proliferate
enormously, and acquire a box-like
shape. In this freeze-fracture electron
micrograph, numerous peroxisomes
(P) are visible, as are the nucleus
(N), the vacuole (V) and
mitochondria (M). × 37 000.
[Courtesy of Dr M. Veenhuis.]

In rat liver the peroxisomes have a characteristic paracrystalline
core, which consists of the enzyme urate oxidase (figure 7.3). However,
the peroxisomes of other animal organs and plant tissues lack this core,
and so they are difficult to distinguish from the many other membrane-
bound vesicles seen in electron micrographs, unless histochemical
staining (for catalase) is used.

One feature of peroxisomes that has proved useful to cell biologists is
that they are *inducible*. Administration of the **hypolipidaemic** drug
Clofibrate to rats leads to a tenfold increase in the number of
peroxisomes per liver cell. In certain yeasts grown on methanol, the
peroxisomes come to occupy almost the whole volume of the cytoplasm
(figure 7.4).

Biochemical analysis of purified peroxisomes has revealed that the
major membrane lipids are phosphatidylcholine and phosphatidylethanolamine, with lesser amounts of cholesterol, sphingomyelin and
phosphatidylinositol. Most of the peroxisomal proteins, including
catalase, are soluble, and so presumably located in the matrix of the
organelle. SDS-PAGE (as in figure 2.8) has identified only three major
integral membrane proteins, with molecular masses of 22, 68, and
70 kDa. The cDNAs for the 22 kDa protein, for catalase and several
other peroxisomal enzymes, have recently (1987) been cloned, so we
should soon know considerably more about their structure and
biosynthesis.

Hypolipidaemic: Causing a
decrease in the concentration of lipid
in the blood.
Clofibrate: Ethylchlorophenoxy-
isobutyrate.

7.2.2   Peroxisomes are involved in lipid degradation and biosynthesis

A β-oxidation pathway, similar in function to that of mitochondria, is present in peroxisomes (figure 7.5). However, the peroxisomal pathway differs from that of mitochondria in several important respects:

(a)   carnitine is not required for the import of the fatty acid;

(b)   it can catalyse the degradation of very-long-chain fatty acids (C20 or greater);

(c)   it does not act on short fatty acids (C6 or less);

(d)   the first enzyme in the pathway is an FAD-linked acyl CoA oxidase, rather than the FAD-linked dehydrogenase of the mitochondrial pathway;

(e)   this first reaction is coupled to the formation of H_2O_2;

(f)   the other enzymes are distinct gene products specific to the peroxisome.

Despite these differences, the mitochondrial and peroxisomal β-oxidation pathways share some common features: the sequence of enzyme-catalysed steps is similar; acetyl CoA is released in each round, leaving a fatty acyl CoA shorter by two carbon atoms; reduced coenzymes ($FADH_2$, NADH) are generated in the process. So what is the relative contribution of each organelle to the degradation of fatty acids by the cell? Estimates vary, but 10–32 per cent of the total oxidation of such compounds may occur in peroxisomes.

The peroxisomal pathway makes two special contributions to lipid degradation. Only it can handle very-long-chain fatty acids, which are particularly prevalent in neuronal glycolipids; and it is also responsible

Figure 7.5
Comparison of β-oxidation in peroxisomes and mitochondria

Fatty acids are degraded in both organelles by the release of acetyl CoA in each turn of the β-oxidation 'spiral'. However, the pathway starts with an oxidase-catalysed reaction in peroxisomes, but with a dehydrogenase-catalysed reaction in mitochondria. The hydratase-acyl CoA dehydrogenase reactions are catalysed by a single bifunctional protein in peroxisomes, but by separate enzymes in mitochondria. The thiolases of the two organelles, like the other β-oxidation enzymes, are distinct gene products. [Taken from Moser, H. W. (1987). *Dev. Neurosci.*, **9**, 1–18; reproduced by permission of the publisher, S. Karger AG, Basel.]

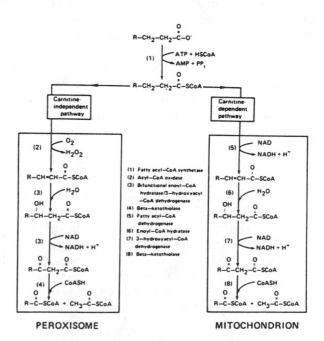

for the degradation of **phytanic acid.** The significance of these features will become apparent when we consider the inherited peroxisomal disorders (section 7.2.4).

Peroxisomes are not merely lipid-degrading organelles. In animals, they also participate in the biosynthesis of certain classes of lipid, specifically the *ether lipids* and (in part) the *bile acids*. The ether lipids (mainly phospholipids) contain a fatty acid linked via an ether bond to the C-1 of the glycerol backbone; in some cases, an adjacent double bond creates a vinyl ether linkage (figure 7.6). The biosynthesis of these compounds involves an unusual exchange of an acyl group on C-1 of a phospholipid with a long-chain fatty alcohol; the pathway is summarised in figure 7.7. The final ether lipid product is transferred to the smooth ER for membrane biosynthesis. Up to 50 per cent of the phosphatidylethanolamine (PE) in brain tissue is of the plasmalogen form, and most tissues contain small but significant amounts of the ether lipids, usually plasmalogen PE. What is the biological role of these lipids? It is not clear, though they are thought to protect cellular membranes from attack by reactive species such as singlet oxygen, generated by oxidative stress.

We do not have the space here to consider the biosynthesis of cholic acid (one of the bile acids) by liver peroxisomes. The reader is referred to the reviews in section 7.6.

7.2.3 Peroxisomes play a special role in plants

At the same time as de Duve was analysing the peroxisomes of animal tissues, several groups of workers were studying an unusual feature of lipid metabolism in germinating seeds, such as those of the castor-oil plant. The endosperm of such seeds is able to bring about **gluconeogenesis** by using fatty acid as a precursor; animal cells conspicuously lack this metabolic capability. The key enzymes in this process are isocitrate lyase and malate synthetase, which are part of the *glyoxylate pathway* (figure 7.8). The net effect of this pathway is the synthesis of one molecule of succinate from two molecules of acetyl CoA. Interestingly, all the enzymes of the glyoxylate pathway were shown to be located in microbody-like structures of the endosperm cell.

Phytanic acid: A long (C20) branched-chain fatty acid derived from the oxidation of the phytol side-chain of chlorophyll.

Gluconeogenesis: Biosynthesis of glucose from non-carbohydrate precursors.

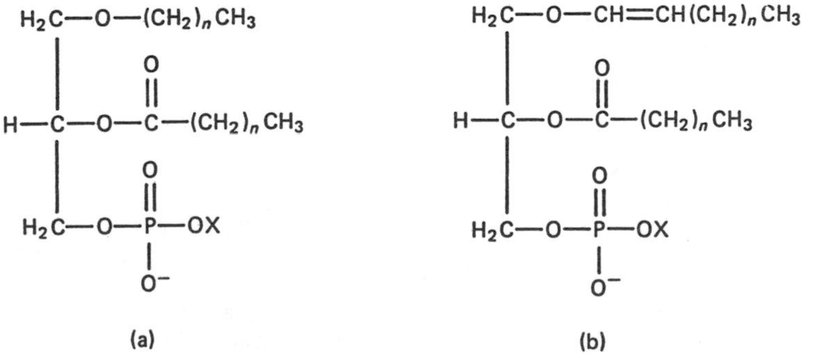

Figure 7.6
Structures of the ether lipids

The acyl group on C-2 is linked *via* the usual ester bond. However, the alkyl group on C-1 is linked *via* an ether bond [in (a)], or a vinyl ether bond [in (b), also called a plasmalogen]. The X group in each phospholipid is predominantly ethanolamine.

Figure 7.7
Biosynthesis of ether lipids

The chief enzymes are: 1, DHAP acyltransferase; 2, alkyl-DHAP synthetase; 3, acyl/alkyl-DHAP reductase. All these enzymes are bound to the membrane of the peroxisome, and have relatively low activity. (DHAP, dihydroxyacetone phosphate.)

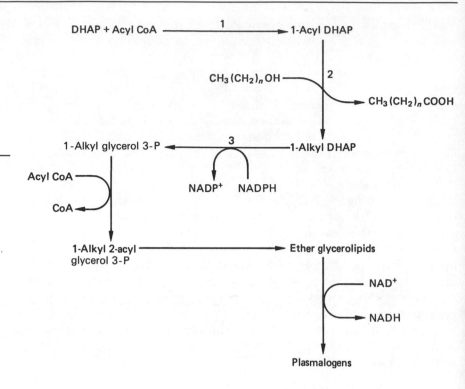

Figure 7.8
The glyoxylate pathway of plant peroxisomes

Although the reactions are shown here as a cycle, labelling studies indicate that most of the malate formed in the peroxisome is exported to the mitochondria. The unique enzymes of the pathway are: 1, isocitrate lyase; 2, malate synthetase. The glyoxylate pathway allows germinating seeds that contain mainly lipid as their food reserve, to convert fat into carbohydrate.

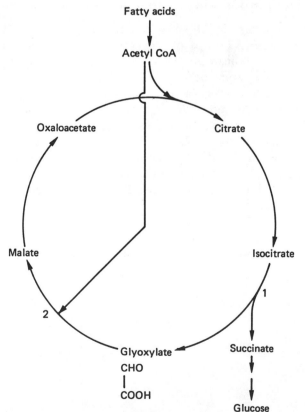

These organelles were thus called *glyoxysomes*. Isocitrate lyase and malate synthetase (but not the other enzymes of the glyoxylate pathway) are also present in peroxisomes of yeast, *Tetrahymena*, *Euglena*, and various fungi.

Interestingly, glyoxysomes are absent from the *leaves* of the castor-oil plant, though the usual type of catalase-containing peroxisome is present. In photosynthetic tissues, the organelle plays a major role in the metabolism of *glycolate*, which is generated by the oxidative process of *photorespiration* in chloroplasts (figure 7.9). Ribulose 1,5-bisphosphate carboxylase, the carbon-fixing enzyme of green plants (figure 9.11), also acts as an oxygenase, so that in many plant species photorespiration may occur at 25–50 per cent of the rate of photosynthesis. The net result is a marked reduction in the maximum possible rate of carbon-fixation in photosynthetic tissues in the light. Thus, peroxisomes take the glycolate produced by chloroplasts and convert it into glycine,

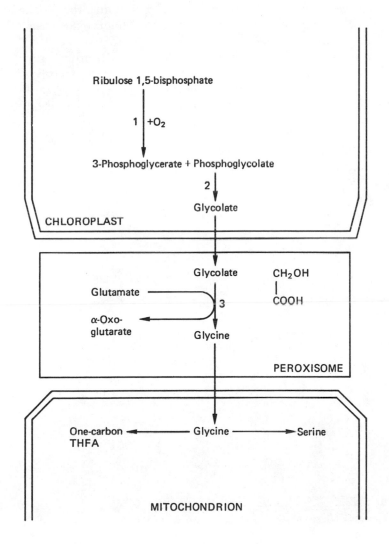

Figure 7.9
The role of peroxisomes in photorespiration

The operation of this photorespiration pathway during photosynthesis reduces the ability of the plant cells to fix CO_2. The principal enzymes are: 1, ribulose bisphosphate carboxylase/oxygenase; 2, phosphoglycolate phosphatase; 3, glycolate transaminase. (THFA, tetrahydrofolic acid, a carrier of activated one-carbon units.)

which is exported to mitochondria for the synthesis of serine or for the formation of one-carbon groups linked to tetrahydrofolic acid. Here is an excellent example of metabolic cooperation between three different organelles (figure 7.10).

7.2.4   Peroxisomal functions are disrupted in certain genetic diseases

In 1973 **Goldfischer** describe the morphological absence of peroxisomes in the liver of a young child who had died of an inherited disease called **Zellweger syndrome.** Here was convincing evidence that peroxisomes do indeed play an essential role in human development. Biochemical studies of *post-mortem* tissue identified several abnormalities: very low levels of ether lipids, and an excessive accumulation of very-long-chain fatty acids, phytanic acid, and some intermediates of bile acid synthesis. Peroxisomal enzymes were deficient or absent; catalase activity was normal, but it was aberrantly located in the cytosol, as determined by *digitonin-permeabilisation experiments* (figure 7.11). Thus, the biochemical abnormalities could be adequately accounted for on the basis of the observed enzyme deficiencies.

Another seven clinically distinct peroxisomal disorders have been described; more will undoubtedly follow. They are currently (1988) classified into three groups on the basis of their aetiology (table 7.1), but this classification of peroxisomal disorders is almost certainly an

Zellweger syndrome: Genetic disorder characterised by severe neuropathology, and malfunction of the liver and kidneys; inherited (usually) in an autosomal recessive manner.

Figure 7.10
Metabolic cooperation between plant organelles

In this leaf cell, a peroxisome is wedged between two chloroplasts (C) and a mitochondrion (M). The peroxisome (P) has a single membrane and a crystalline core of catalase. × 41 000. [Taken from Frederick, S. E. and Newcomb, E. H. (1969). *J. Cell Biol.*, **43**, 343–353; reproduced by copyright permission of the Rockefeller University Press.]

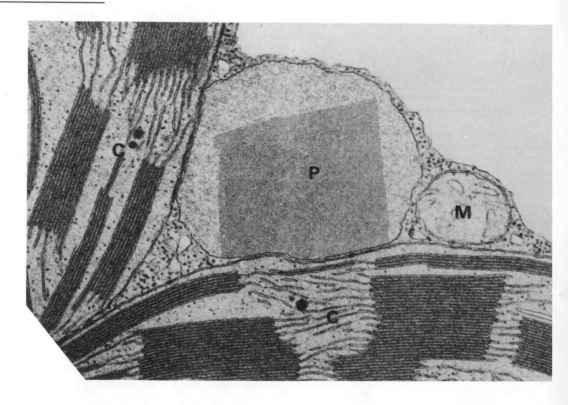

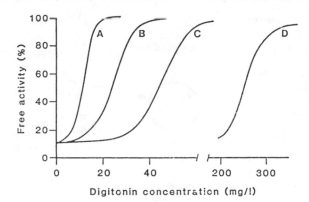

Figure 7.11
Digitonin-permeabilisation studies

Fibroblast cells were incubated under isotonic conditions in the presence of increasing concentrations of the detergent, digitonin. The permeabilisation (disruption) of the plasma membrane (curve A) was assessed by measuring the release of the cytosolic marker enzyme, lactate dehydrogenase; of lysosomal membranes (curve B) by the release of hexosaminidase; of mitochondrial membranes (curve C) by the release of glutamate dehydrogenase; and of peroxisomal membranes (curve D) by the release of catalase. In cells from a patient with Zellweger syndrome, the release of catalase follows curve A, rather than curve D. This result suggests that in this disorder, the soluble peroxisomal enzymes are aberrantly located in the cytosol.

over-simplification. **Complementation analysis** with cells from several unrelated patients with Zellweger syndrome suggest the existence of at least three genetically distinct forms of the disease. The nature of the genetic lesions in these peroxisomal disorders is also likely to be disparate. Pulse-chase experiments (figure 2.10) indicate that peroxisomal enzymes are synthesised normally in patients with Zellweger syndrome, but they fail to become localised within peroxisomes. The enzymes presumably undergo degradation by cytosolic proteases. At least one X-linked peroxisomal disorder is known, and so its underlying genetic lesion must differ from that of the autosomal recessive conditions. The cloning and sequencing studies currently being applied to peroxisomal enzymes should also permit the precise delineation of the mutation in individual patients with a peroxisomal disorder.

Complementation analysis: Two patients, X and Y, suffering from apparently the same inherited disease, may have allelic mutations (in the same gene) or non-allelic mutations (in different genes). If cells from patients X and Y are fused, there are two possible outcomes: (1) the hybrid cells may display the original biochemical defect, in which case the parent cells probably have allelic mutations; (2) the hybrid cells may have normal metabolism, when the parent cells are said to have corrected or *complemented* each other, probably as a result of the mutations being non-allelic.

7.3 Biogenesis of peroxisomes

The current model for the biogenesis of peroxisomes is that proposed by **Lazarow**; its key features are presented in figure 7.12. This model accounts nicely for several important aspects of peroxisomal structure and function, and makes experimentally testable predictions for other aspects. Any model must take into account the fact that:

(1) all peroxisomal proteins (including the integral membrane proteins) are synthesised on free ribosomes in the cytosol;

(2) peroxisomal proteins are not glycosylated;

Table 7.1
Features of some peroxisomal disorders

Group	Representative example	Peroxisomes	Peroxisomal enzymes
I	Zellweger syndrome	Absent or deficient	Deficient or normal (but cytosolic)
II	Pseudo-Zellweger syndrome	Present in normal numbers	Several enzymes deficient
III	Acatalasaemia	Present in normal numbers	Single enzyme deficient

Figure 7.12
**Model for the biogenesis of
peroxisomes**

Proteins synthesised on free ribosomes
are incorporated post-translationally
into pre-existing peroxisomes. Fission
of an enlarged organelle generates
daughter peroxisomes.

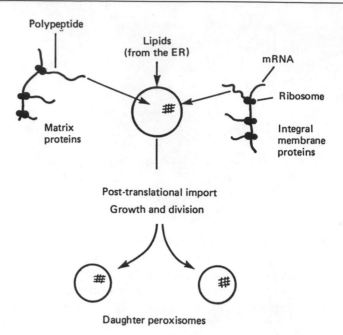

(3)   peroxisomes lack DNA;
(4)   in certain yeast species (*Candida*), electron microscopy has
       revealed peroxisomes apparently undergoing a process of division;
(5)   the matrix of the peroxisome is mildly acidic (about pH 6 in
       Candida).

Lazarow proposes that peroxisomes grow and undergo fission to form
new peroxisomes; in this respect, they would thus resemble mitochondria
and chloroplasts in their mode of biogenesis. The early ideas of Novikoff,
that the smooth ER played a key role in generating new peroxisomes,
have now been largely discredited. The only contribution of the
endoplasmic reticulum in the current model is in the biosynthesis of
the lipids of the peroxisomal membrane.

One prediction of Lazarow's model (figure 7.12) is that peroxisomal
proteins synthesised on free ribosomes are imported into the organelle
post-translationally. This prediction has been amply confirmed. Newly
synthesised urate oxidase labelled *in vivo* in the presence of ^{14}C-leucine
was incorporated within a few minutes into peroxisomes. Thus, this
process differs fundamentally from the co-translational translocation
and processing that accompany the biosynthesis of secretory proteins
(as described in section 4.2.2). What are the features of a nascent
peroxisomal protein that direct its incorporation into a peroxisome?
We do not yet know. The proteins are invariably synthesised as the
mature polypeptide, rather than as a larger precursor. There is no
cleavable amino acid sequence at the *N*-terminal end that would act
as a 'signal peptide', as is present in secretory, mitochondrial,
and chloroplast proteins. However, information on this point should
soon be forthcoming. The process of importing proteins into peroxisomes

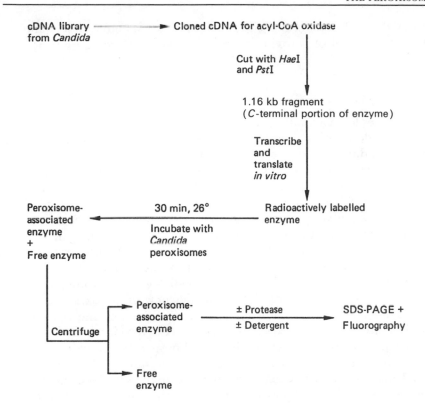

Figure 7.13
**Reconstitution of protein
import by peroxisomes**

Radioactive acyl-CoA oxidase,
synthesised from the cDNA *in vitro*,
was taken up by peroxisomes isolated
from the yeast, *Candida*. The extent of
protein import was assessed by
resistance to added protease in the
absence of added detergent. (In the
presence of detergent, added protease
completely degrades peroxisome-
associated enzyme.)

of *Candida* has been reconstituted *in vitro* (figure 7.13). The major finding
of these studies was that the information needed to internalise acyl-CoA
oxidase into peroxisomes resided in the *C*-terminal portion of the
protein. Here again was another unique feature of the peroxisomal
import process.

Complementation analysis of peroxisomal disorders (section 7.2.4)
suggests that at least five distinct gene products are required for the
biogenesis of peroxisomes. These may be proteins regulating the
expression of genes for peroxisomal enzymes; or integral polypeptides
of the peroxisomal membrane; or as yet uncharacterised cytosolic
factors. A similar experimental approach has exploited mutant CHO
cells (from Chinese hamster ovary) which are deficient in peroxisomes.
No complementation is observed between different cell lines, so they
all presumably have the same allelic mutation. **Revertant cells** have
not been observed. This result is consistent with a 'seeding' requirement
for the biogenesis of peroxisomes: if the mutant cells lack these
organelles, then there will be no progenitor peroxisome from which
daughter organelles can arise, even if the original genetic mutation is
corrected. Further analysis of this system may lead to the identification
of the various factors required for the biogenesis of peroxisomes.

So how did the peroxisome originate in the course of evolution?
Did it evolve as a result of an association between a prokaryotic cell
and a primordial eukaryotic cell — the '*endosymbiont hypothesis*'? If so,
it has lost all of the DNA of the prokaryote, and it no longer has the

Revertant cells: Mutant cells
which have regained their original
function, usually as a result of a
second, corrective, mutation in the
same gene.

expected double membrane (figure 1.7). Perhaps a better understanding of the expression of peroxisomal proteins will shed some light on this puzzle.

7.4 Summary

Peroxisomes form a family of ubiquitous, catalase-containing organelles, with a single limiting membrane and rather variable morphology. As well as oxidative enzymes which give rise to H_2O_2, they contain a unique β-oxidation pathway for the degradation of very-long-chain fatty acids. Peroxisomal enzymes are also involved in the biosynthesis of ether lipids and bile acids in animals. In plants, specialised peroxisomes (glyoxysomes) convert fatty acids to precursors for gluconeogenesis *via* the glyoxylate pathway. In photosynthetic tissues, peroxisomes cooperate with chloroplasts and mitochondria in the process of photorespiration. The essential role of peroxisomes in human development is vividly illustrated by Zellweger syndrome, a lethal genetic disease characterised by a deficiency of peroxisomes. New peroxisomes arise by growth and division of pre-existing organelles. Newly synthesised peroxisomal proteins leave cytosolic ribosomes as the mature polypeptides, which are imported post-translationally *via* an as yet uncharacterised mechanism; likewise, the nature of the targeting signal on nascent peroxisomal proteins is unknown. Long neglected, the peroxisome is now known to make a significant contribution to the metabolism of most cells.

7.5 Study questions

1. How can peroxisomes be distinguished from lysosomes in thin sections of tissue examined by electron microscopy? (section 7.2.1)
2. How might the fact that rat liver peroxisomes are induced by Clofibrate (section 7.2.1), assist in the cloning of cDNA for peroxisomal proteins?
3. What other proteins (other than those mentioned in chapter 7) might be present in the peroxisomal membrane?
4. In the Group II peroxisomal disorders (table 7.1), patients may synthesise 'empty' peroxisomes which have a normal membrane, but lack the soluble enzymes of the organelle's matrix. How could such vesicles be identified in thin sections of pathological tissue?
5. In plant leaves, what might be the consequences of a deficiency of peroxisomes? (section 7.2.3)

7.6 Further reading

de Duve, C. (1983). *Sci. Amer.*, **248** (5), 74–84.
   (Historical overview)
Tolbert, N. E. (1981). *Ann. Rev. Biochem.*, **50**, 133–157.
   (Peroxisomal metabolism)

Borst, P. (1983). *Trends Biochem. Sci.*, **8**, 269–272.
  (Peroxisomes of animals re-assessed)
Lazarow, P. B. and Fujiki, Y. (1985). *Ann. Rev. Cell Biol.*, **1**, 489–530.
  (Biogenesis of peroxisomes)
Moser, H. W. (1987). *Dev. Neurosci.*, **9**, 1–18.
  (Peroxisomal disorders)

8 THE MITOCHONDRION

8.1 Introduction

Mitochondria have long been one of the most studied subcellular organelles. Why is this? Firstly, their size (up to 5 μm long) and characteristic morphology meant that they were readily identified by microscopists. Secondly, they house numerous important reactions of oxidative metabolism, in particular those leading to the synthesis of ATP. Thirdly, they contain all the genetic machinery required to express several mitochondrial components, and thus have a degree of autonomy from the nucleus (a characteristic shared with chloroplasts, section 9.5). In this text, we cannot consider in detail the morphology of mitochondria, nor the numerous reactions of intermediary metabolism that occur there; these aspects are well covered in standard biochemistry textbooks. Instead, we shall focus on three important areas of research concerning mitochondrial structure and function: the generation of ATP by oxidative phosphorylation; the genetic machinery of these organelles; and their biogenesis and evolution. First, however, we need to set the scene by outlining the contribution that the mitochondrion makes to cellular function.

8.1.1 Morphology and distribution of mitochondria

Mitochondria are found in all eukaryotic cells, with the exception of mammalian erythrocytes and certain primitive Protozoa such as *Microsporidia*. Under the electron microscope (figure 8.1), they appear as circular or oval profiles, which represent spheres 0.5–2 μm in diameter or cylinders 0.2 μm across by up to 5 μm in length, respectively. The number of mitochondria per cell varies enormously, from one up to several thousands; in a typical mammalian cell (table 1.1) there may be several hundred. In general terms, the concentration of mitochondria can be roughly correlated with the cell's demand for metabolic energy in the form of ATP. The organelles are often localised within the cell at the sites of energy consumption (figure 8.2).

The images of this organelle as presented in electron micrographs may give a misleading impression of a discrete, static structure. In

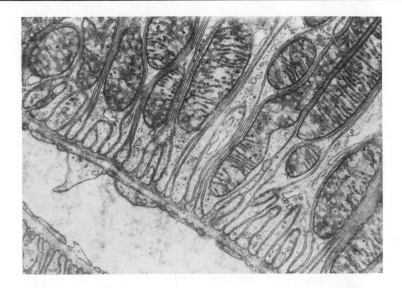

Figure 8.1
Morphology of mitochondria

In this thin section through a kidney cell, the mitochondria have well-developed arrays of internal membranes. They lie alongside infoldings of the plasma membrane, where they provide the metabolic energy (in the form of ATP) required for active transport. [Taken from Fawcett, D. W. (1981). *The Cell*, 2nd edn, W. B. Saunders Co., Philadelphia. Original micrograph courtesy of A. Ichikawa.]

reality, within the cell, the mitochondrion is a dynamic organelle that probably has a complex morphology. Time-lapse cinematography of cells examined by phase-contrast microscopy has revealed marked changes in the size and shape of individual mitochondria. When serial sections of yeast cells were viewed under the electron microscope, there was evidence for mitochondria having extensively branched structures much larger than previously observed.

When ultra-thin sections of cells are examined at very high magnification (figure 8.3), the characteristic ultrastructure of the mitochondrion becomes readily apparent. The *cristae* of the highly folded inner membrane provide an enormous surface area for membrane-associated events, such as those involved in electron transport and

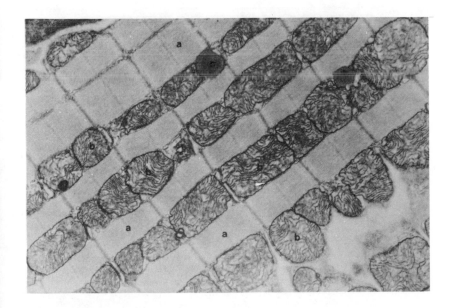

Figure 8.2
Intracellular distribution of mitochondria

Mitochondria are often present at sites of high energy demand within the cell. In this thin section of cardiac muscle, numerous mitochondria (b) lie between the myofibrils (a). The cristae are particularly well developed, consistent with this tissue's high demand for ATP during contraction. One source of the metabolic energy is lipid, present in the form of lipid droplets (c) in the cytoplasm. [Taken from Macleod, A. G. (1975). *Cytology*, The Upjohn Company, Kalamazoo. Original micrograph courtesy of K. R. Porter.]

Figure 8.3
Ultrastructure of the
mitochondrion

(a) In this longitudinal section
through a mitochondrion of a
pancreatic cell, the outer membrane
(at b) and inner membrane (at c)
surrounding the organelle are
apparent. The cristae (a) are seen to
arise from infoldings of the inner
membrane (for example, at d).
Several electron-dense granules (e)
are present. (b) This schematic
representation shows the relationship
between the two mitochondrial
membranes and the two spaces they
enclose. (c) At high magnification of
negatively-stained preparations, the
inner surface of the cristae is seen to
be decorated with 'lollipop'-shaped
structures (arrow) with
approximately spherical heads
9–10 nm in diameter; these are the
site of ATP synthesis. [(a) courtesy of
D. W. Fawcett. (c) taken from
Fernandez-Moran, H. *et al.* (1964).
J. Cell Biol., **22**, 63–100; reproduced
by copyright permission of the
Rockefeller University Press.]

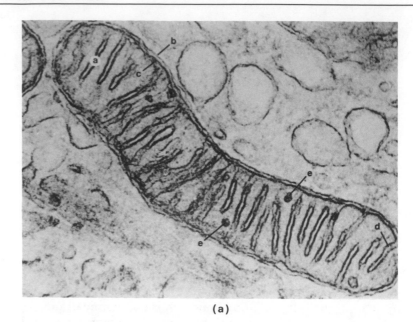

(a)

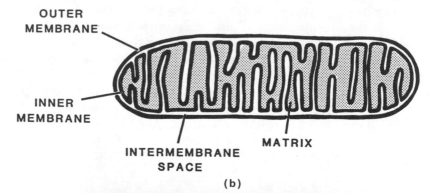

(b)

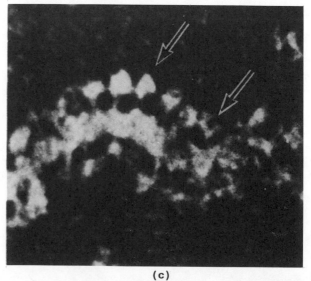

(c)

oxidative phosphorylation (section 8.2). Indeed, the number of cristae per mitochondrion is in direct proportion to the cell's requirement for ATP. The apparently structureless *matrix* of the organelle may be seen to contain dense granules of uncertain composition, DNA, and ribosomes bound to the inner membrane. Bear in mind that the ultrastructure of mitochondria can vary enormously from that depicted in figure 8.3a, not only between different organisms, but also within one organism under different physiological conditions.

The lipid composition of the outer mitochondrial membrane is unexceptional (table 1.5). The inner membrane, which consists of approximately 70 per cent protein and 30 per cent lipid, has phosphatidylcholine and phosphatidylethanolamine as the major lipids; in comparison with other subcellular membranes, cholesterol is a relatively minor component. The inner membrane also has an unique phospholipid that forms about 15 per cent of the total: *cardiolipin,* or diphosphatidylglycerol (figure 8.4). We shall consider the proteins of the mitochondrial membranes in relation to their function (sections 8.2 to 8.4).

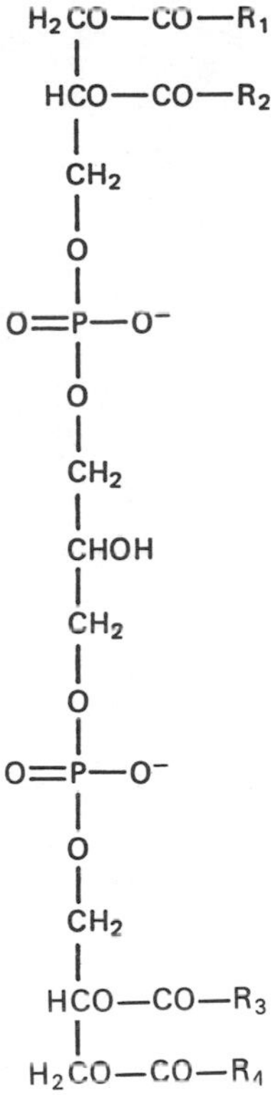

8.1.2   The mitochondrion's contribution to metabolism

It is a relatively simple matter to prepare mitochondria from tissue homogenates (by differential and density-gradient centrifugation; section 2.2.1), and from there to 'dissect' the two membranes and two spaces of the organelle by detergent extraction or sonication. Biochemical characterisation of these four components, in terms of protein and enzyme composition, allows one to assign discrete metabolic functions to them (table 8.1). Apart from the provision of ATP, the metabolic role of the mitochondrion includes the following: the complete oxidation of carbohydrates, fats and the carbon skeletons of amino acids (to CO_2

Figure 8.4
Structure of cardiolipin (**diphosphatidylglycerol**)

R_{1-4} are the alkyl chains of the fatty acids. This phospholipid also occurs in the plasma membrane of bacteria.

Compartment	Enzyme or metabolic activity
Outer membrane	1. Synthesis of cardiolipin
	2. Monoamine oxidase
Inter-membrane space	1. Adenylate kinase
	2. Cytochrome *c* peroxidase
Inner membrane	1. Succinate dehydrogenase (Krebs cycle)
	2. Electron-transferring flavoprotein of β-oxidation
	3. Carnitine acyltransferase
	4. Electron transport chain
	5. ATP synthase
Matrix	1. Krebs (tricarboxylic acid) cycle
	2. β-Oxidation of fatty acids
	3. Ketone body metabolism
	4. Haem synthesis
	5. Some reactions of urea synthesis and gluconeogenesis

Table 8.1
Metabolism in mitochondrial compartments

The table is not exhaustive; representative examples only are shown.

and H_2O); the provision of oxaloacetate for gluconeogenesis; the synthesis and degradation of ketone bodies (such as 3-hydroxybutyrate); the synthesis of urea (in the liver of ureotelic organisms); and the synthesis of haem and (in part) steroid hormones. Thus, the mitochondrion should be considered as having a major biosynthetic function within the cell, as well as a degradative one.

The metabolism within the mitochondrion does not take place in isolation from that in the rest of the cell. We have already seen (section 1.3.2) how metabolites are shuttled into and out of the organelle in order to link metabolic pathways (or parts of a pathway) that occur in different subcellular sites. The inner mitochondrial membrane is relatively impermeable to most solutes, which can only be transported if the appropriate permease is present. There are numerous such transport proteins or *translocases* (table 8.2). The concentration gradients involved, or thermodynamic constraints, invariably mean that solutes pass in only one direction across the inner membrane. In contrast, the outer membrane is relatively permeable to low-molecular-mass solutes. It contains a 30 kDa protein called *porin*, oligomers of which create non-specific channels permeable *in vitro* to oligosaccharides of up to 8 kDa.

8.2 'Powerhouse of the cell'

In 1948 **Palade** described a procedure (a form of differential centrifugation, in fact) for isolating morphologically intact mitochondria.

**Table 8.2
Transport of mitochondrial metabolites**

Translocase	Function IN	OUT	Metabolic role
Phosphate	$H_2PO_4^-$	OH^-	Mitochondrial ATP synthesis
Adenine nucleotide	ADP^{3-}	ATP^{4-}	Mitochondrial ATP synthesis
Pyruvate	$Pyruvate^-$	OH^-	Link between glycolysis and Krebs (TCA) cycle
Dicarboxylate	$Malate^{2-}$	HPO_4^{2-}	Transfer of reducing equivalents
Tricarboxylate	$2\text{-Oxoglutarate}^{2-}$	$Citrate^{3-} + H^+$	1. Export of acetyl CoA 2. Control of glycolysis 3. Replenishment of cytosolic NADPH
2-Oxoglutarate	$2\text{-Oxoglutarate}^{2-}$	$Malate^{2-}$	Transfer of reducing equivalents
Glutamate–aspartate	$Glutamate^{2-} + H^+$	$Aspartate^{2-}$	Transfer of reducing equivalents

The translocases (○) are located in the inner mitochondrial membrane.
(IN = mitochondrial matrix, OUT = inter-membrane space)

Soon afterwards, **Lehninger** showed that mitochondria isolated by this method catalysed the complete oxidation of fatty acids. Thus, the organelle must contain all the enzymes associated with β-oxidation and the Krebs cycle, as well as the electron transport system. Lehninger also demonstrated that mitochondria could exploit the free energy derived from electron transport from NADH to O_2 in order to drive the synthesis of ATP from ADP and P_i, in the process of respiratory chain-linked oxidative phosphorylation. These experiments led to the concept of mitochondria as the 'powerhouse' of aerobic cells. They also represented some of the first evidence that the metabolic functions of the cell might be localised in a specific subcellular organelle. How is the oxidation of reduced coenzymes coupled to the synthesis of ATP?

8.2.1 Reduced coenzymes are oxidised *via* the electron transport chain

The pioneering studies of **Keilin, Chance,** and many other workers led to the identification of the components of the mitochondrial electron transport chain, and their likely sequence. **Green, Hatefi** and co-workers were able to purify these constituents by chromatography of detergent-solubilised mitochondria. They found that functionally related components were associated in macromolecular aggregates or *complexes*. Some of the properties of Complexes I to IV are presented in table 8.3. By mild sonication of mitochondria, **Racker** was able to prepare *sub-mitochondrial particles*, whose components had the opposite orientation to that in native mitochondria (figure 8.5). By means of non-penetrating probes such as specific antibodies, he and others were able to determine the orientation within the inner mitochondrial membrane of the components of the electron transport chain. They are currently thought to be organised as shown in figure 8.6.

Table 8.3
Composition of the electron transport complexes

Complex[a]	Molecular mass ($\times 10^3$ kDa)[b]	Polypeptides	mtDNA-encoded polypeptides	Prosthetic groups	Ratio in mitochondria[c]
I	0.7–0.9	25	7	FMN, FeS clusters	1
II	0.14	4–5	—	FAD, FeS clusters, b_{560} haem	2
III	0.25	9–10	1	b_{562}, b_{566}, c_1 haems, [2Fe–2S] cluster	3
IV	0.16–0.17	8	3	aa_3 haems, Cu_A Cu_B	6–7
V	0.5	12–14	2	Adenine nucleotides	3–5

Data are those for mitochondria from bovine heart.
[a] Systematic names for these complexes are: I, NADH: ubiquinone oxidoreductase; II, succinate: ubiquinone oxidoreductase; III, ubiquinol: ferricytochrome c oxidoreductase; IV, ferrocytochrome c: oxygen oxidoreductase (cytochrome oxidase); V, ATP synthase (F_0F_1-ATPase).
[b] Polypeptide component only.
[c] Based on the content (one molecule per complex) of FMN (Complex I), covalently-bound FAD (Complex II), cytochrome c_1 (Complex III), cytochrome aa_3 (Complex IV), and F_1 component (Complex V).

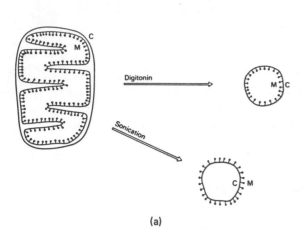

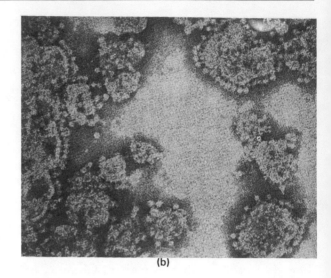

(a)   **(b)**

Figure 8.5
Preparation of
submitochondrial particles

(a) Particles with normal orientation
are prepared by treatment with the
detergent, digitonin; particles with
inverted orientation are prepared by
sonication. C = cytosol, M = matrix.
(b) Electron micrograph of negatively
stained submitochondrial particles
prepared by sonication. The F_1 heads
of the F_0F_1-ATPase can be clearly
seen projecting from the membrane.
[(a) taken from Whittaker, P. A. and
Danks, S. M. (1978). *Mitochondria:
Structure, Function and Assembly*,
Longman, London; reproduced by
permission of the publisher.
(b) courtesy of E. Racker.]

Redox potential: A physico-
chemical measure of the tendency of a
system to donate or accept electrons.
Electrons are passed from a system
with a relatively high negative redox
potential (such as $NADH/NAD^+$) to
a system with a more positive redox
potential (such as O_2/H_2O). See
figure 9.4 for an example of how
electrons are passed from one redox
component to another.

Inhibitors have provided useful clues as to the organisation of the
electron transport machinery (table 8.4). With a measuring device for
following respiration, such as an *oxygen electrode* (figure 8.7), and by
using appropriate substrates, and inhibitors, it has been possible to
'dissect' the components of the electron transport chain and their
interactions.

From a consideration of the **redox potentials** of the components
of the four complexes, it was considered likely that there were three
sites where there was a sufficient decrease in free energy to drive the
endergonic synthesis of ATP: at Complexes I, III, and IV. Thus, the
transfer of a pair of electrons from NADH to O_2 could generate three
molecules of ATP; each $FADH_2$ could give rise to only 2 ATP, since
this reduced coenzyme donates electrons to Complex II, thereby
by-passing the energy-conserving step of Complex I. These values came
to be known as *P/O ratios*, as each pair of electrons led to the reduction
of one atom of oxygen. The P/O ratios of 3 (for NADH) and 2 (for
$FADH_2$) were important landmarks that had to be accounted for in
any model of the mechanism of oxidative phosphorylation, as we shall
see (section 8.2.2). Electron transport coupled to ATP synthesis is not
restricted to mitochondria: functionally similar systems occur in the
plasma membrane of prokaryotes, and in the chloroplasts of photosynthetic organisms (as will be discussed in section 9.3).

The detailed structure of many of the components of the
mitochondrial electron transport chain is now known. In part, such
information has been derived from classical protein chemistry. For
example, the peripheral membrane protein *cytochrome* c was relatively
easily solubilised and purified to homogeneity; the amino acid sequence
of the pure protein was determined, and its three-dimensional
conformation derived from X-ray crystallographic studies. However,

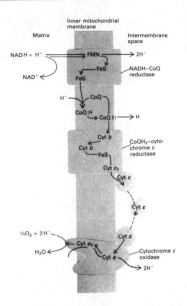

Figure 8.6
Organisation of the electron transport chain

The components of the electron transport chain (ETC) are thought to be arranged in the inner mitochondrial membrane as shown in this diagrammatic representation. The transfer of electrons along the ETC is coupled to the vectorial expulsion of protons from the mitochondrial matrix. Note how Complexes I, III and IV all span the inner membrane. (FeS, iron–sulphur protein; Cyt, cytochrome.) [Taken from Darnell, J., Lodish, H. and Baltimore, D. (1986). *Molecular Cell Biology*; reproduced by copyright permission of Scientific American Books Inc.]

it has proved extremely difficult to apply the same procedures to some of the highly hydrophobic polypeptides, such as cytochrome *b*. Nevertheless, it has been possible to predict the structure of such components from the amino acid sequence derived from the base sequence of cloned cDNA. The overall size and shape of Complexes III and IV have been determined at low resolution by electron microscopy (with image enhancement) of two-dimensional crystals.

The main conclusion of these studies is that Complexes I, III and IV form polypeptide aggregates which span the inner mitochondrial membrane (as depicted in figure 8.6). Their components are orientated such that electrons take a zig-zag path, traversing the membrane several times. **Ubiquinone** and cytochrome *c* act as mobile electron carriers linking the transmembrane complexes.

The possible arrangement of the 25 or so polypeptides of Complex I is depicted in figure 8.9. The *iron–sulphur (FeS) proteins* play an essential role in the process of electron transfer from NADH to ubiquinone; yeast mutants deficient in FeS proteins are unable to respire normally.

Ubiquinone: A polyprenoid lipid capable of accepting or donating either one or two hydrogen atoms. Its structure is depicted in figure 8.8. Its hydrophobic nature allows it to diffuse freely in the lipid bilayer of the inner mitochondrial membrane. Also known as Coenzyme Q, CoQ.

Table 8.4
Effect of inhibitors on mitochondrial electron transport

Inhibitor	Site of block in electron transfer
Rotenone	Complex I $\rightarrow$ CoQ
Phenolytrifluoroacetone	Complex II $\rightarrow$ CoQ
Antimycin A	Complex III $\rightarrow$ Cytochrome *c*
CN^-, CO, N_3^-	Complex IV $\rightarrow O_2$

Only representative examples are shown. Electron carriers which act prior to the site of inhibition become completely reduced; those which act after the site of inhibition become totally oxidised. Oligomycin blocks electron transport by inhibiting the F_0 component of the mitochondrial ATP synthase. CoQ, coenzyme Q (ubiquinone).

Figure 8.7
Mitochondrial respiration studied with the oxygen electrode

(a) In the Clark oxygen electrode, O_2 is reduced to H_2O at the platinum (Pt) electrode. The resulting current is proportional to the O_2 concentration in the solution. A typical chamber contains 3 ml of medium, and requires 3–5 mg of mitochondrial protein. (b) In these representative traces showing the oxidation of succinate (succ) by mitochondria (mitos), O_2 concentration is the ordinate and time the abscissa. In the presence of medium alone (lower trace), added ADP stimulates respiration but the effect is blocked by oligomycin (inhibits ATP synthase). In the presence of the uncoupler carbonyl cyanide-*p*-trifluoromethoxy-phenylhydrazone (FCCP), rapid respiration is blocked by cyanide (inhibits electron transport) (middle trace) or by n-butylmalonate (inhibits import of succinate) (upper trace). [Taken from Nicholls, D. G. (1982). *Bioenergetics*, Academic Press, London; reproduced by permission of the publisher.]

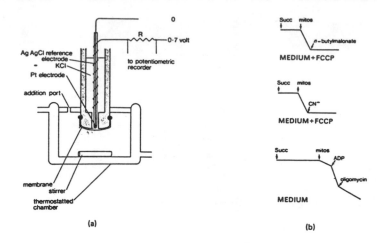

(a)

(b)

The major component of Complex II is *succinate dehydrogenase*, a membrane-bound enzyme of the Krebs cycle that contains covalently bound FAD. The bovine enzyme consists of two subunits (70 and 27 kDa) non-covalently bound to two hydrophobic polypeptides (15.5 and 13.5 kDa) associated with a **cytochrome b** (b_{560}). The latter is structurally and genetically distinct from the b cytochromes (b_{562}, b_{566}) present in Complex III. Complex III itself consists of 9–10 polypeptides associated with three redox centres (cytochromes b_{562}, b_{566}, and c_1). The composition of the complex from the mould *Neurospora crassa*, and its possible organisation in the membrane, are presented in table 8.5 and figure 8.10 respectively. The presence of two b cytochromes with differing spectral properties might point to two distinct polypeptides. In fact, there is a single gene for the polypeptide encoded in the mitochondrial DNA; presumably, the two proteins are in slightly different environments. Cytochromes b and c_1, as well as the FeS protein, have been sequenced; they all contain at least one hydrophobic segment which probably represents the membrane attachment site. Complex III

Cytochrome b_{560}: One of a family of b cytochromes, with a wavelength of maximum light absorption at 560 nm. (Compare cytochrome $P450$; section 4.3.2)

Figure 8.8
Structure of ubiquinone (Coenzyme Q)

The structure shown is that of the oxidised form of CoQ found in mammals. In its reduced form (ubiquinol), the two quinoid oxygen atoms appear as hydroxyl groups. In the plastoquinones of plants, the two methoxy groups are replaced by methyl groups.

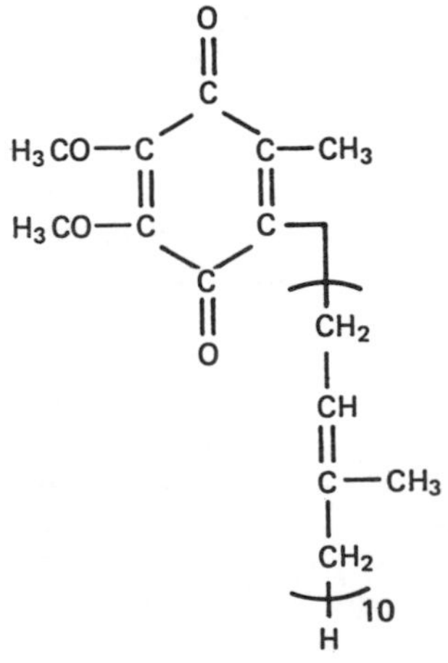

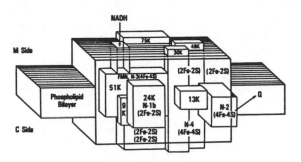

Figure 8.9
Organisation of Complex I

The possible orientation of the
components of Complex I is shown
here diagrammatically. NADH binds
to the 51 kDa subunit and donates
electrons to the matrix face (M side)
of the complex. These electrons are
passed ultimately to ubiquinone (Q)
via polypeptides that contain
iron–sulphur clusters (FeS) and the
flavin-containing cofactor, FMN. The
lipid-free complex from the fungus,
N. crassa, has a molecular mass of
about 610 kDa, and can be resolved
by SDS-PAGE into 24 different
polypeptides. Recent (1987) data
indicate that about two-thirds of the
complex projects 10 nm away from
the inner mitochondrial membrane on
the M side. [Taken from Hatefi, Y.
(1985). *Ann. Rev. Biochem.,* **54,**
1015–1069; reproduced, with
permission, copyright 1985 by Annual
Reviews Inc.]

of mitochondria is structurally and functionally homologous to the cytochrome b/f complex of plant chloroplasts (figure 9.4). Complex IV, or *cytochrome oxidase,* is a multi-subunit aggregate which catalyses the transfer of electrons from reduced cytochrome c to molecular O_2. Its composition is presented in table 8.6. Long stretches of hydrophobic amino acids are present in all the subunits except V, and such sequences probably anchor the proteins in the inner mitochondrial membrane (figure 8.11). Subunit VI is thought to be exposed on the matrix face of the inner mitochondrial membrane, whereas subunit V probably lies at the cytosolic face. One copper ion (called Cu_A, or Cu_a) is associated with a β-pleated sheet structure in subunit II, the other (Cu_B, or Cu_{a_3}) with a transmembrane α-helical segment of subunit I, as are cytochromes a and a_3. Oxygen binds between cytochrome a_3 and Cu_B. The sequence of electron transfer within Complex IV is thought to be: reduced cytochrome $c \rightarrow$ cytochrome $aCu_A \rightarrow$ cytochrome $a_3Cu_B \rightarrow O_2$. How does a molecule of oxygen acquire four electrons from cytochrome c? *Electron spin resonance* studies suggest an intermediate in which an O^{2-} species is complexed to a cytochrome aCu_A that has given up one electron each from the haem group ($Fe^{2+} \rightarrow Fe^{3+}$) and the copper ion ($Cu^+ \rightarrow Cu^{2+}$). Acquisition of two further electrons and

Polypeptide	Molecular mass (kDa)	Number of copies	Polypeptide	Molecular mass (kDa)	Number of copies
1 (core protein 1)	50	1	6	14	1
2 (core protein 2)	45	2	7	12	1
3 (cytochrome b)	42[a]	1	8	11	1
4 (cytochrome c_1)	29[a]	1	9	8	1
5 (FeS protein)	22[a]	1			

Table 8.5
Polypeptide composition of Complex III

The data refer to Complex III from the fungus *N. crassa.* Other eukaryotes have a complex of closely comparable composition. Cardiolipin (diphosphatidylglycerol; figure 8.4) is probably also associated with Complex III in the inner mitochondrial membrane. The entire lipid-free complex has a molecular mass of about 550 kDa, and so exists as a dimer *in vivo.*

[a] Molecular mass calculated from DNA sequence data. Other values are apparent molecular masses estimated by SDS-PAGE.

Figure 8.10
Organisation of Complex III

In this schematic representation, the dashed line outlines the cytochrome bc_1 subcomplex. The numbers refer to the molecular masses of core proteins 1 (50 kDa) and 2 (45 kDa) in the complex from the mould, *N. crassa*. Calculations of the distribution of hydrophobic residues indicate that cytochrome b contains 8–9 membrane-spanning segments; cytochrome c_1 has one such sequence at its *C*-terminal end. The haem group of cytochrome b is coordinated to 4 histidine residues in membrane-spanning segments II and V. FeS, iron–sulphur protein. [Adapted from Li, Y. *et al.* (1981). *FEBS Lett.*, **135**, 277–280; reproduced by permission of Elsevier Science Publishers.]

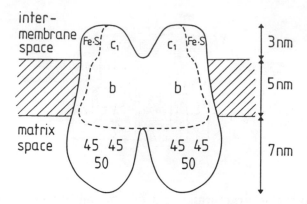

four protons generates two molecules of H_2O at the matrix face of the inner membrane. In conclusion, we now know the identity of the 70 or so different polypeptides of the mitochondrial electron transport machinery, but we are not yet able to define the functions of all of them.

8.2.2   A proton gradient drives ATP synthesis

By the late 1950s, the identity of the major components of the electron transport chain was well established, as was the P/O ratio for the oxidation of NADH and $FADH_2$. What was not understood was the nature of the mechanism whereby electron transfer was *coupled* to the synthesis of ATP from ADP and P_i. The likely identity of the enzyme which catalyses this process was known: it was the ATP synthase, more commonly called the F_0F_1-*ATPase*. Negatively stained preparations of mitochondria examined by electron microscopy revealed knob-like projections on the matrix face of the inner membrane (see, for example, figure 8.3c). When submitochondrial particles were treated with mild

Table 8.6
Composition of Complex IV

Polypeptide	Molecular mass (Da)	Membrane-spanning segments[a]
I	56 993	12
II	26 049	2
III	29 918	6
IV	17 153	1
V	12 436	0
VI	9 419	1
VII-Ser	5 541	1
VII-Ile	4 962	1
VII-Phe	6 244	1

The composition shown is one research group's data for Complex IV (cytochrome oxidase) from bovine heart. However, there is considerable disagreement among researchers as to the identity of the polypeptide subunits of Complex IV and their precise stoichiometry.

[a] Calculated on the basis of each segment being a sequence of about 20 non-polar amino acids capable of forming a transmembrane α-helix.

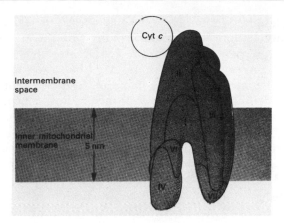

Intermembrane space

Cyt c

Inner mitochondrial membrane 5 nm

Figure 8.11
Organisation of Complex IV

The components of Complex IV (cytochrome oxidase) are thought to be oriented in the inner mitochondrial membrane as shown in this diagrammatic representation. It is not known if Complex IV exists *in vivo* as a monomer or a dimer of this aggregate of subunits. Cytochrome *c* (Cyt *c*) binds mainly to subunit II. For full activity, cytochrome oxidase requires the presence of tightly bound lipid, probably including cardiolipin (figure 8.4). [Adapted from Brunori, M. and Wilson, M. T. (1982). *Trends Biochem. Sci.*, **7**, 295–299; reproduced by permission of Elsevier Science Publications.]

denaturing agents, the spherical end-piece of each projection was released from the membrane in a water-soluble form. This F_1 protein component was associated with the stalk and an integral membrane protein structure called F_0. Together, the F_0F_1 complex *in vitro* acted as an ATP-hydrolysing enzyme (hence the name); in intact mitochondria, it formed the ATP synthase, which exploited the free energy released by electron transport to drive the synthesis of ATP. But what was the mechanism whereby these two processes were coupled together during oxidative phosphorylation?

In the mid-1960s there were three main hypotheses to account for oxidative phosphorylation: the chemical coupling model of **Slater**, first proposed in 1953; the conformational coupling model of **Boyer** (1964); and the chemiosmotic coupling model of **Mitchell** (1961). It is a matter of historical fact that Mitchell's concept of chemiosmosis has best withstood the test of time and experiment, in one of the most hotly disputed areas of biochemical research. We should note, however, that elements of Boyer's concepts may prove useful in defining in detail the events which occur at the F_1 component of the ATP synthase during oxidative phosphorylation, as discussed below.

The principles of the chemiosmotic theory are as follows:

1. Transfer of electrons *via* the electron transport chain leads to the vectorial extrusion of protons from the mitochondrial matrix into the inter-membrane space.
2. The electrochemical potential of the proton gradient so created drives the endergonic synthesis of ATP by the F_0F_1–ATPase.
3. The lipid bilayer of the inner mitochondrial membrane is impermeable to protons.
4. The components of the electron transport chain are asymmetrically arranged in the inner mitochondrial membrane.
5. Uncouplers have their effect by transferring protons across the inner membrane back into the matrix, thereby collapsing the proton gradient.

Although there is little dispute over these broad concepts, there is considerable disagreement over the precise details of the mechanism of oxidative phosphorylation. For example, what is the exact *stoichiometry* of proton translocation (that is, how many protons are pumped out for each electron pair passing from NADH to O_2)? What is the mechanism of proton translocation? Are the translocated protons delocalised in the inter-membrane space, or do they remain associated with the inner membrane? How does the inward flow of protons through the F_0 component of the ATP synthase energise the formation and release of ATP by the F_1 component?

Even Mitchell has modified his ideas in order to account for certain features of oxidative phosphorylation revealed by subsequent experiments. His original postulate of six protons expelled per pair of electrons transferred ($6H^+/2e^-$) is now considered a gross under-estimate; a figure of 12 or 13 is thought to be more likely. His original model did not assign a role to cytochrome oxidase in the process of proton translocation. However, purified Complex IV does transport protons vectorially when incorporated into liposomes. Such a finding is more consistent with a 'proton-pump' mechanism of proton extrusion, rather than with Mitchell's original 'redox-loop' model, with its rather rigid stoichiometry. His proposed value of $6H^+/2e^-$ was consistent with two protons being imported per ATP synthesised. However, a proton is also imported when ATP^{4-} in the matrix is exchanged in an electro-neutral process for ADP^{3-} and HPO_4^{2-} in the inter-membrane space (table 8.2). The proton gradient is also used to drive the mitochondrion's accumulation of Ca^{2+}, and the transfer of metabolites across the inner mitochondrial membrane (table 8.2).

Whatever the precise mechanism of proton translocation, the free energy of the electrochemical proton gradient is exploited by the ATP synthase. The F_1 component of the enzyme from animal mitochondria, as well as bacteria and plant chloroplasts, has an unusual subunit stoichiometry: $\alpha_3\beta_3\gamma\delta\varepsilon$ (table 8.7). F_1 from rat liver mitochondria has been studied by X-ray crystallography to 0.9 nm resolution. A unique $\alpha\beta$ pair makes contact with a complex of γ, δ and ε on one side, and

Table 8.7 **Composition of the mitochondrial ATP synthase**	Subunit	Molecular mass (kDa)	Subunit	Molecular mass (kDa)
	α	56	F_6	9.0
	β	53	IF_1	9.6
	γ	33	F_B	11–12
	δ	16	6	23
	ε	11	8	5.9
	OSCP	21	9	11

The data refer to the enzyme (total mass 500 kDa) from bovine heart. The enzyme from bacteria (*E. coli*) has polypeptides homologous to subunits α, β, γ, δ and 6, so these proteins probably arose at an early stage of evolution. OSCP, oligomycin-sensitivity conferring protein, is present in two copies per F_1 unit. The proteolipid subunit 9 of F_0 is present in 6 copies per F_1 unit.

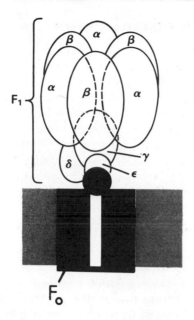

Figure 8.12
Organisation of
mitochondrial ATP synthase
(F_0F_1-ATPase)

The catalytically active portion of the aggregate, situated on the matrix face of the inner mitochondrial membrane, comprises a trimer of ($\alpha\beta$) subunits. This structure, which forms an asymmetric flattened ellipsoid, corresponds to the head of the 'lollipops' visible in electron micrographs, such as figure 8.3c. It is connected by the γ and δ subunits to a membrane-bound F_0 component. The function of the F_1 ε subunit is not known. [Adapted from Walker, J. E., Saraste, M. and Gay, N. J. (1982). *Nature*, **298**, 867–869; reproduced by permission of Macmillan Magazines Ltd]

with the remaining α and β subunits on the other side (figure 8.12). ATP analogues label the β subunit of this unique $\alpha\beta$ pair and thereby identify it as the presumed catalytic subunit. What is the function of the remaining two $\alpha\beta$ pairs? Does each of the three $\alpha\beta$ pairs take part in turn in ATP synthesis by means of a 'rotational catalysis' mechanism? This recently proposed model needs to be tested. Sequence homology between the F_1-β subunit and another enzyme that catalyses the synthesis of ATP, *adenylate kinase*, reveals a flexible polypeptide loop which closes over the MgATP molecule. Several amino acid residues in the loop are potential proton acceptors. Perhaps Boyer's concept of a 'conformational coupling' mechanism contains an element of truth: the free energy of the proton gradient may indeed be required to release tightly bound ATP from the catalytic site of the ATP synthase.

The 'stalk' which connects the F_1 complex to the F_0 component of the ATP synthase is composed of two proteins: a 21 kDa polypeptide (OSCP) which binds a potent inhibitor of ATP synthesis, *oligomycin*; and a 9 kDa polypeptide, coupling factor 6 (F_6). The role of these proteins in proton translocation from F_0 to F_1 is unknown, as is that of the ATPase inhibitor protein, IF_1, and coupling factor F_B.

The F_0 component of the mammalian ATP synthase consists of at least three different polypeptides (table 8.7), specified by genes encoded by mitochondrial DNA. The role of subunits 6 and 8 is unclear, but subunit 9 has long been known as the **proteolipid** which binds *dicyclohexylcarbodiimide* (DCCD). This inhibitor forms a covalent bond with glutamate 58 of the proteolipid, and in so doing blocks proton translocation. Site-directed mutagenesis may tell us about the functions of individual amino acid residues in the components of F_0, and thus about the mechanism of coupling between the proton gradient and ATP synthesis that occurs at the interface of F_0 and F_1.

Proteolipid: Polypeptide with a significant amount of covalently bound lipid.

8.3   Mitochondrial genetics

We have already mentioned (section 8.1) that mitochondria have their own genetic machinery, with DNA and ribosomes that can support the synthesis of some of the organelle's own proteins. Nevertheless, most mitochondrial proteins are encoded by nuclear genes. How is the genetic information of the mitochondrion organised? How is its expression coordinated with that of the nuclear genes? What is the effect of mutations in the mitochondrial genome? These and other questions are pertinent not only to our consideration here of mitochondrial genetics, but also in relation to the biogenesis and evolution of this organelle (section 8.4).

8.3.1   Mitochondria have an unique genetic system

We have already seen (table 3.1) how the mitochondria of Man and yeast do not use the standard set of codons of the genetic code. This is only one of several features which distinguish the genetic machinery of the mitochondrion from that of the nucleus. What are the others?

DNA was first isolated from chick mitochondria in 1966. Under the electron microscope, it appeared as a supercoiled, covalently closed circle (figure 8.13). Remarkably, it was devoid of the histone-containing nucleosome structures which characterise nuclear DNA (section 3.4.1), and in this respect, the mitochondrial genome more closely resembles that of prokaryotes, with their naked circular double helix. However, most mitochondrial DNA is much smaller than that of bacteria (table 8.8), in fact much more akin to plasmids.

The relatively small size of the human mitochondrial genome assisted the determination of its base sequence, reported by **Sanger**'s group in 1981. The organisation of this mitochondrial DNA is extremely

Figure 8.13
Mitochondrial DNA

This circular DNA molecule from a mitochondrion of a *Xenopus* oocyte has a contour length of about 5 μm. It clearly lacks the nucleosome structure of chromosomal DNA, as depicted in figure 3.14. Not visible in this electron micrograph is the 'D-loop', a region of partially triple-helical DNA that contains the origin of replication of one of the strands. [Taken from Fawcett, D. W. (1981). *The Cell*, 2nd edn, W. B. Saunders Co., Philadelphia; courtesy of I. Dawid.]

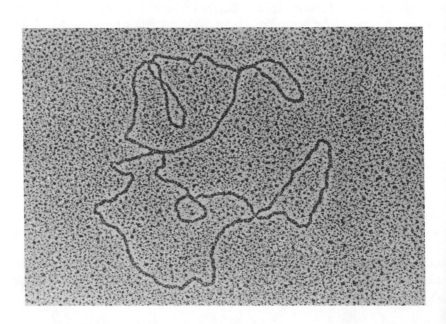

Source	Form	Length (kb)	**Table 8.8**
Human	Circular	16.569	**Structure of mitochondrial**
Yeast	Circular	75	**genomes**
Plant	Circular	250–2400	
Tetrahymena	Linear	50	
(Bacterium	Circular	4000)	

For comparison, the structure of the DNA of the prokaryote *E. coli* is given. The ciliated protozoon, *Tetrahymena*, is one of the few eukaryotes with linear mitochondrial DNA. The nuclear DNA of yeast comprises 13 500 kb, 180 times larger than that of its mitochondrial DNA.

economical: almost all of its sequence, of both its strands, is transcribed and translated (figure 8.14). The coding regions specify 2 ribosomal RNA species (16S and 12S), 22 transfer RNA species, and 5 integral membrane proteins. At the time of its determination, the sequence also contained 8 **unassigned reading frames**, URFs. The identity of most of these URFs has since been determined; seven of them code for hydrophobic polypeptides of Complex I. All the protein-coding genes are present on one strand of the double helix, invariably flanked by a tRNA gene at each end. Other tRNA genes are dispersed on the complementary strand. All the genes are flush-ended, that is, there are no intergenic non-coding sequences; one consequence of this fact is that there is an apparent lack of regulatory elements which might control transcription of the mitochondrial genes. Introns are lacking,

Unassigned reading frames: Sequence of DNA which, if transcribed and translated, would specify an uncharacterised polypeptide.

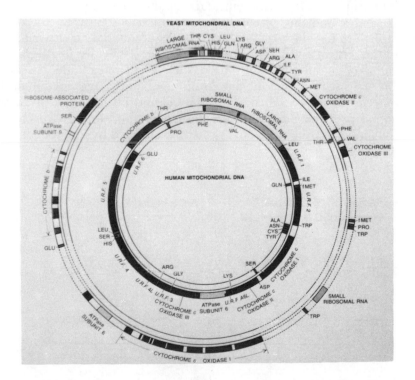

Figure 8.14
Maps of the mitochondrial genome

The positions of known genes are indicated on the two inner circles (human) and two outer circles (yeast). Broken lines in the latter represent unsequenced segments. The tRNA genes are given their amino acid abbreviations (appendix A). Note that the much larger yeast mitochondrial genome contains introns in several genes and many non-coding sequences. U.R.F., unassigned reading frame (now known to code for subunits of Complex I). [Taken from Grivell, L. A. (1983). *Sci. Amer.*, **248**, (3), 60–73; copyright © 1983 by Scientific American, Inc. All rights reserved.]

and there is only one reading frame for each strand. The rRNA and tRNA species are smaller than their nuclear-encoded counterparts.

By comparison, the mitochondrial DNA of the yeast *S. cerevisiae* is much less compact (figure 8.14). Most of its genes have been identified, and it has introns in several genes (for example, that for cytochrome *b*, subunit I of cytochrome oxidase, and the 21S rRNA). Furthermore, the introns of the gene for cytochrome *b* contain coding sequences which specify '*maturases*' required for RNA splicing. Such a process generates in some cases small circular RNA species, which may be 'lariat' intermediates of the type formed by splicing of nuclear pre-mRNA.

Transcription of the two strands of the human mitochondrial DNA yields two long polycistronic precursor RNAs. How is each mature mRNA and rRNA generated from this precursor? It appears that an endonuclease, encoded by a nuclear gene, cuts the polycistronic RNA precisely at the ends of each mRNA, rRNA, and tRNA species. Since there are no intergenic segments that might act as recognition sequences, the endonuclease presumably recognises the folded conformation of the tRNA which flanks the mRNA and rRNA sequences. The mitochondrial mRNAs have two unusual features. Firstly, unlike nucleus-encoded mRNAs, they lack 5' leader sequences that might act as ribosome-binding sites for the initiation of translation; it is not clear how mitochondrial ribosomes recognise the AUG initiation codon. (Furthermore, AUA and AUU can also act as start signals.) Secondly, some mitochondrial mRNAs lack a termination codon. However, polyadenylation at the 3' end generates such a codon (figure 8.15).

The mitochondrial ribosomes are smaller than their cytoplasmic counterparts, and thus resemble prokaryotic ribosomes. Furthermore, like the latter, they are inhibited by antibacterial agents such as chloramphenicol and tetracycline. In contrast, they are unaffected by inhibitors of cytoplasmic translation, such as cycloheximide. Assembly of ribosomes within the mitochondrion involves the aggregation of the two species of rRNA, encoded by mitochondrial genes, with all the ribosomal proteins, encoded by nuclear genes and imported from the cytosol (in yeast, one ribosomal protein is specified by a mitochondrial

Figure 8.15
Novel creation of stop codons in mitochondrial mRNA

The primary transcripts of several human mitochondrial genes lack conventional stop codons. After endonuclease-catalysed cleavage, addition of adenine nucleotides to the 3' end generates a translation–termination sequence (boxed).

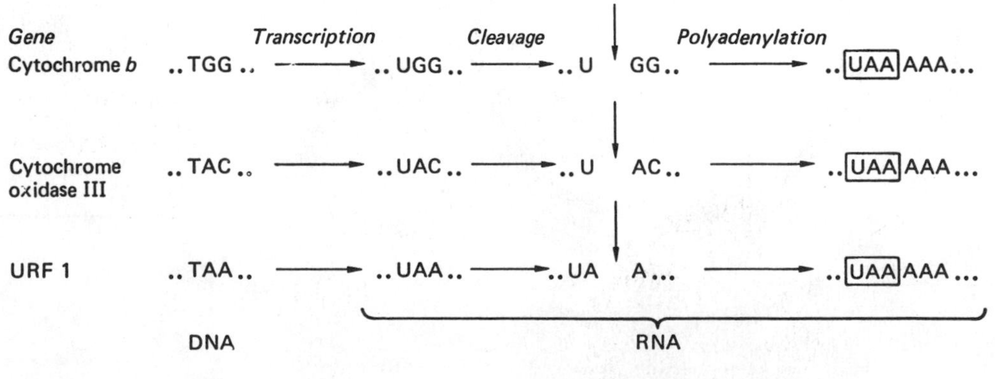

Gene	Transcription	Cleavage	Polyadenylation
Cytochrome *b*	..TGG.. ──→ ..UGG.. ──→ ..U GG.. ──→ ..⟦UAA⟧AAA...
Cytochrome oxidase III	..TAC.. ──→ ..UAC.. ──→ ..U AC.. ──→ ..⟦UAA⟧AAA...
URF 1	..TAA.. ──→ ..UAA.. ──→ ..UA A... ──→ ..⟦UAA⟧AAA...

DNA   RNA

gene). The 61 codons in the mitochondrial mRNA are read by only 22 tRNAs (in humans), far fewer than the minimal set of 32 predicted by **Crick.** What is the explanation? It seems that each of the tRNAs can read a 'family' of either two or four synonymous codons. In the two-codon families, 'wobble' at the third base means that G can pair with U as well as with C. In the four-codon families, there appears to be no Watson–Crick type of pairing between the third bases; only the first two bases of the codon are recognised. This total lack of specificity at the third base is not seen in the corresponding cytoplasmic system.

8.3.2 Mitochondrial mutants have proved informative

That mitochondria might harbour their own genetic machinery was deduced from a study of the inheritance of certain mutant characteristics in yeast. This organism has proved to be extremely useful in the study of mitochondrial genetics. It is easily cultured, but more importantly it has the ability to survive by means of either aerobic or anaerobic metabolism. Thus, lesions which incapacitate the mitochondria are not lethal in these yeast mutants, as they would be in most other cells.

The original discovery of mitochondrial mutants in yeast was made by **Ephrussi** in 1949. He found organisms which grew slowly, and thereby formed smaller colonies than normal, when cultured on agar plates containing glucose as the carbon source. Two of the three classes of these *petite* (*pet*) mutants had a genetic lesion which was not present in the nuclear DNA. Later analysis revealed that the 'neutral *petites*' had completely lost their mitochondrial DNA, whereas the 'suppressive' *petites* had a roughly normal amount of mitochondrial DNA but it had a grossly abnormal composition. (Some rare *petite* mutants had genetic lesions in their nuclear DNA.) Analysis of similar yeast mutants led to the identification of several families of mutations in the mitochondrial DNA (table 8.9). In fact, conventional genetic studies led to the

Class	Complementation group	Gene(s) affected	Gene product(s) affected
pet	ρ^-, ρ^0	Many	Cytochromes a, a_3, b; rRNAs
mit$^-$	COX	*oxi* 1, 2, 3	Subunits of Complex IV
	BOX	*box*	Cytochrome *b*
	COX/BOX	*oxi + box*	Cytochrome *b* + Complex IV
	PHO	*pho* 1, 2	Subunits of ATP synthase
—	[bI]	Introns in *box*	Maturase?
syn$^-$	—	*var* 1	Ribosomal protein
	—	lgRNA	21S rRNA
	—	*asp*, etc.	tRNAs

Table 8.9

Mutations in mitochondrial DNA of yeast

The ρ^- (rho minus) and ρ^0 (rho zero) mutants correspond to the suppressive *petites* and neutral *petites*, respectively. Data refer to the yeast *S. cerevisiae*.

mapping of many of the mitochondrial genes long before their older was deduced from the nucleotide sequence of the DNA. Mitochondria from certain mit^- yeast mutants have proved extremely valuable in studies of the biogenesis of the organelle (section 8.4).

Nuclear *pet* mutants of yeast, induced experimentally by mutagenesis, are also proving informative in studies of the regulation of expression of mitochondrial genes and the assembly of the respiratory chain-associated complexes. Complementation analysis (section 7.2.4) indicates that there may be as many as 200 nuclear genes which influence the synthesis of mitochondrial components. The properties of the nuclear *pet* mutants are summarised in table 8.10.

Since most of the mitochondria of the zygote are contributed by the ovum in organisms which use sexual reproduction, the maternal inheritance of mitochondrial genes (and thus mutations) does not display typical Mendelian characteristics.

In the complex mitochondrial genome of higher plants, the DNA may exist as small linear molecules in addition to the more usual circular forms. The *master circle* DNA of maize (with a total mitochondrial DNA of 570 kb) contains six repeated elements. These homologous sequences allow recombination events to take place between them, thereby changing the orientation of some genes or generating small subgenomic circles, the latter visible under the electron microscope. Certain mutant forms of *Petunia* species display **cytoplasmic male sterility** (CMS). Cloning and sequencing of the region of mitochondrial DNA associated with CMS revealed a 1062 bp open reading frame. This abnormal gene (*Pcf*) contains elements of the genes for subunit 9 of the ATP synthase and subunit II of the cytochrome oxidase, as well as an unassigned reading frame. Antibodies raised to synthetic peptides predicted from the base sequence locate a 25 kDa CMS-associated polypeptide present only in the anthers of mutant plants. This 25 kDa protein in some way interferes with mitochondrial biogenesis, but we do not yet know how.

Cytoplasmic male sterility: Affected plants produce no pollen, because the tapetal layer of cells in the anther decays at a very early stage of development.

Table 8.10
Nuclear *pet* mutants of yeast

	None	Multiple	Enzyme deficiency F_0F_1-ATPase	Complex III	Complex IV
Number of complementation groups	80	45–50	3–4	40	30
Gene product(s) identified	1. Enzymes of CoQ biosynthesis 2. Expression of cytochrome *c*	1. Aminoacyl-tRNA synthetases 2. Ribosomal proteins 3. Elongation factor	1. α subunit of F_1 2. β subunit of F_1	1. 5′-end processing of cyt. *b* pre-mRNA 2. Maturase 3. Translation of cyt. *b*	1. Subunit IV 2. Subunit V 3. Subunit VI 4. Maturases

Each complementation group represents a separate gene. cyt., cytochrome. Data refer to the yeast *S. cerevisiae*.

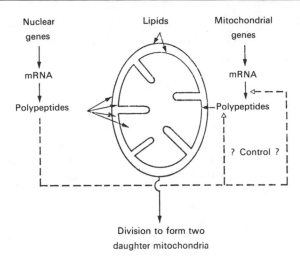

Division to form two
daughter mitochondria

Figure 8.16
Biogenesis of mitochondria

Structural genes in the nuclear DNA
give rise to proteins incorporated into
the two spaces and two membranes of
the mitochondrion. Mitochondrially
encoded proteins are components of
the inner membrane only. The broken
lines indicate that nuclear genes exert
some form of control over the
expression of mitochondrial genes.
(The mitochondrial genes for tRNA
and rRNA are not included here.)

8.4   Biogenesis and evolution of mitochondria

Mitochondria synthesise their own rRNA and tRNA, and a small
number of proteins, but the large majority of the organelle's polypeptides
have to be imported from the cytosol after their synthesis by free
ribosomes. This situation immediately poses some intriguing questions
concerning the biogenesis of mitochondria (figure 8.16). How do
proteins synthesised by cytoplasmic ribosomes gain access to the
mitochondrion? How do such proteins become localised in the
appropriate compartment of the organelle? How is the expression of
the nuclear genes for mitochondrial polypeptides coordinated with that
of the mitochondrial genes? Since new mitochondria arise by division
of an enlarged pre-existing organelle (figure 8.17), there must be some

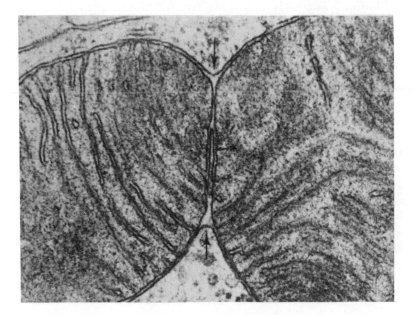

Figure 8.17
**Mitochondrion undergoing
division**

The outer membrane has started to
form a cleft at top and bottom
(arrows), and a slender invagination
has appeared (asterisk). The inner
membrane has already completed this
process. [Taken from Fawcett, D. W.
(1981). *The Cell*, 2nd edn,
W. B. Saunders Co., Philadelphia.
Original micrograph courtesy of T.
Kanaseki.]

Replication of mitochondrial DNA
Mitochondrial DNA contains a short triple-helix structure, the 'D-loop', which arises from displacement of the parental heavy (H) strand by a nascent H strand of about 0.6 kb, stably associated with the closed circular duplex. Replication starts with expansion of the D-loop, displacing the parental H strand. When about 2/3rds of the H strand has been replicated in a clockwise direction, the origin of light (L) strand synthesis is exposed. Mitochondrial replication involves an organelle-specific primase (an RNA polymerase, which requires for activity a 5.8S rRNA species imported from the cytosol), and a γ-DNA polymerase. (The D-loop also contains promoters for transcription of the H and L strands.)

mechanism for replicating the mitochondrial genome, as well as for the actual division process. These are some of the questions we address in this section.

8.4.1 Most mitochondrial proteins are imported post-translationally

The synthesis of cytochrome c by cultured hepatocytes is inhibited by cycloheximide, which blocks cytoplasmic translation but not mitochondrial translation. Radioactive cytochrome c, synthesised in a cell-free translation system, is taken up by isolated mitochondria in a manner that makes it resistant to added protease. (Compare the synthesis *in vitro* of secretory proteins, section 4.2.1.) The association between the labelled protein and the organelle is a saturable process, which can be blocked by adding excess unlabelled cytochrome c. These experimental observations have led to the concept of a receptor-mediated import of mitochondrial proteins that occurs post-translationally (at least, for the majority of such proteins).

The general features of this process are now clear. They are associated with two major events: (a) binding of a nascent mitochondrial protein to a specific receptor on the outer membrane, with the subsequent translocation of the protein across the membrane; (b) processing of the translocated protein and its integration into the appropriate mitochondrial compartment. Let us consider each of these aspects in turn. (The proteins of the outer mitochondrial membrane do not follow this route. They associate with receptors and integrate directly into the membrane in the absence of any energy requirement.)

Apoprotein: Polypeptide portion of a protein which contains a prosthetic group or other non-protein component(s). For example, functional cytochrome c consists of the apoprotein (apocytochrome c) associated with a haem group.

(a) *Recognition and translocation*: The outer membrane of mitochondria from *Neurospora crassa* has a protease-sensitive component which binds the **apoprotein** form of cytochrome, c, but not cytochrome c itself. The binding process is not inhibited by added cytochrome c_1, so presumably these two proteins have their own specific receptors. The bound apocytochrome c is rapidly transferred to the inter-membrane space, where it acquires its haem group. The translocation is blocked *in vitro* by adding a non-functional haem analogue. This result demonstrates that translocation is not obligatorily coupled to receptor binding.

What features of the nascent mitochondrial protein are recognised by the corresponding receptor? Most mitochondrial proteins are synthesised as larger precursors with amino-terminal extensions of 25–50 amino acids (table 8.11). A survey of several such precursor proteins has revealed no obvious sequence homologies between them, though they do all contain an excess of positively charged and hydroxyl-containing amino acids. Likewise, there is no common sequence around the site where the extension is cleaved by a protease in the mitochondrial matrix. However, projection of the pre-sequence in the form of a 'helical wheel' generates an *amphilic helix*, which may insert itself into the membrane as shown in figure 8.18. A synthetic peptide, constructed only from arginine, serine, and leucine and having

Protein	Location	Amino-terminal pre-sequence

Alcohol dehydrogenase	Matrix	`+ +    ++      +   +`
		`MLRTSSLFTRRVQPSLFSRNILRLQST..`
		`1         10        20`

Subunit IV of cytochrome oxidase	Inner membrane	`+  +  +  +   +      +`
		`MLSLRQSIRFFKPATRTLCSSRYLL..`
		`1         10    ▲  20      ▲`

Cytochrome c_1	Inter-membrane space	`+    ++    +  +` `MFSNLSKRWAQRTLSKSFYSTATG`

Table 8.11
Targeting of mitochondrial proteins

Cytochrome c_1 continued:

```
+    ++    +  +
MFSNLSKRWAQRTLSKSFYSTATG
1         10        20  |
                        |
 ┌──────────┬───────────┘
 │IGAAAVGATVL│ + + +  |
 │    40     │KQTLKGSKSAA
 │           │    30
 │TASTLLYA DSLTAEA ▲ MTA..
 │    50        60
 └──────────┘
```

70 kDa protein | Outer membrane

```
++    + +
MKSFITRNK ┌─────────────┐
1       10│TAILATVAATGTAIG
          │      20     |
          │             |
          │             |
   ++ +   │             |
 ..KKGR   │QQQQQLQNYYYYA
   40     │     30
          └─────────────┘
```

The amino acid sequences (abbreviations in appendix A) are annotated with positively charged residues (+), known sites of proteolytic cleavage (▲), and putative membrane-spanning segments (boxed). Note also the unusually high proportion of hydroxyl-containing amino acids, serine (S), and threonine (T). Additional cleavage sites are known to exist for alcohol dehydrogenase (between residues 25 and 27) and for cytochrome c_1 (between residues 22 and 30).

an amphiphilic helical structure, was fused to the *N*-terminal of a cytosolic enzyme, dihydrofolate reductase (DHFR). The chimaeric protein was readily taken up by isolated mitochondria into a protease-resistant form. The binding between the modified DHFR and the mitochondrion was blocked by a monoclonal antibody directed against a 45 kDa protein of the outer membrane. Perhaps this antibody can be used to characterise the import machinery. The association between the amphiphilic helix and the receptor requires a membrane potential (negative inside), but not the hydrolysis of ATP.

Translocation of the bound protein requires that the polypeptide chain be first unfolded in an ATP-dependent process. (Import of the chimaeric DHFR was blocked if the normal conformation of the enzyme was established by binding of the active-site inhibitor, methotrexate.) It is likely that a specific enzyme ('*unfoldase*') catalyses this rate-limiting step. Subsequent translocation, which is independent of ATP or an electrochemical potential, is probably driven by the free energy of re-folding of the polypeptide chain. What is the precise site at which translocation takes place? Experimental evidence suggests that import

Figure 8.18
Model for translocation of mitochondrial proteins

(a) The diagram relates to the first 18 residues of the precursor of subunit IV of yeast cytochrome oxidase (for amino acid sequence, see table 8.11). If an α-helical model of the sequence is constructed, then a view down the axis of the helix reveals a characteristic disposition of amino acids. Five positively charged groups are clustered at the top right (including the amino group of the N-terminal methionine, M), whereas all the remaining residues are hydrophobic (underlined) or uncharged. 85 per cent of mitochondrial pre-sequences have this 'amphiphilic helix' structure. (Amino acid abbreviations are given in appendix A.) (b) As proposed by **Schatz** and colleagues, the exposed N-terminal pre-sequence forms an amphiphilic helix which associates with the outer leaflet of the lipid bilayer. Reorganisation of some phospholipid molecules forms a partial 'inverted micelle', which stabilises insertion of the pre-sequence into the mitochondrial membrane. Translocation then proceeds.

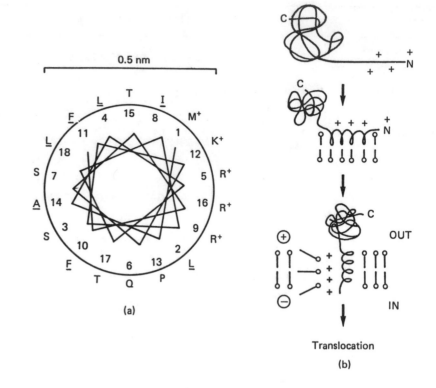

(a)

Translocation

(b)

occurs at regions where there is close contact between the outer and inner mitochondrial membranes (figure 8.19).

(b) *Processing and localisation*: Import of mitochondrial proteins exposes their N-terminal pre-sequences to the matrix. Here, cleavage by a specific endoprotease releases the pre-sequence, a process which prevents the loss of the precursor protein from the organelle. In fact, this proteolytic processing occurs after translocation is complete and the polypeptide has re-folded. Some mitochondrial proteins pass completely across the contact sites and come to lie free in the matrix (for example, enzymes of the Krebs cycle), or they become associated with the matrix face of the inner membrane (for example, F_1 subunits). Other proteins have a hydrophobic 'stop-transfer' sequence which fixes them in the inner membrane. (See section 4.4 for general discussion of the integration of membrane proteins.)

Cytochrome c is an extrinsic protein attached to the cytoplasmic face of the inner mitochondrial membrane. It lacks a pre-sequence, and is transported as the apoprotein only across the outer membrane. In the inter-membrane space, the acquisition of a haem group brings about a conformational change in the protein and thereby maintains its desired localisation. By contrast, nascent cytochrome c_1 follows a circuitous route. The apoprotein precursor is processed by the matrix protease, and the subsequent covalent attachment of the haem group is followed by a second proteolytic cleavage by an enzyme in the

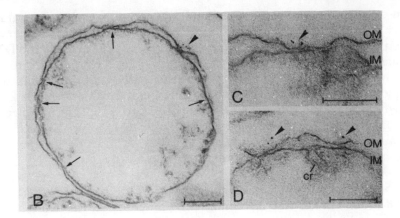

Figure 8.19
Protein import at mitochondrial 'contact sites'

Isolated mitochondria, swollen in hypotonic buffer, were incubated with the F_1 β subunit of the ATP synthase, complexed with antibody (IgG). (Under these conditions, a translocational intermediate is formed, spanning the membranes with the IgG disposed externally). The IgG can be visualised under the electron microscope when it has formed a complex with gold-labelled Protein A, added subsequently. (B) Arrows indicate unlabelled sites of close contact between outer and inner membranes; arrowhead points to a contact site labelled with Protein A–gold. (C) As (B), but higher magnification. (D) As (C), but before fixation the mitochondria were subjected to shearing forces. The outer membrane is ruptured on each side of the contact sites. OM, outer membrane; IM, inner membrane; cr, crista. Bars = 0.2 μm. [Taken from Schwaiger, M., Herzog, V. and Neupert, W. (1987). *J. Cell Biol.*, **105**, 235–246; reproduced by copyright permission of the Rockefeller University Press.]

inter-membrane space. A 'stop-transfer' sequence anchors cytochrome c_1 in the inner membrane (table 8.11).

Thus, the incorporation of a newly synthesised mitochondrial protein into the correct compartment of the organelle with the correct orientation is probably achieved by its having appropriate 'targeting sequences'. However, one should not forget that many mitochondrial proteins lack an *N*-terminal pre-sequence, and some which do have one, lack the usual positively charged amphiphilic helix.

8.4.2   How is mitochondrial gene expression controlled?

The expression of nuclear genes specifying mitochondrial proteins must be coordinated with that of mitochondrial genes. The nature of this interaction also depends on the environment. For example, during anaerobic fermentation by yeast in glucose-containing medium, there is a dramatic decrease in the rate of mitochondrial protein synthesis, but there is no evidence for transcriptional control. Furthermore, in mammalian mitochondria the highly compact genome contains no DNA sequences which might act as promoters or other control elements regulating mRNA synthesis.

Nuclear controller genes have been identified from studies of *pet* mutants of yeast. For example, PET 54, 55 and 494 are three cell lines with defective expression of subunit III of cytochrome oxidase (COX III). Although different gene loci are involved, the yeast mutants have the same phenotype, and Northern blotting reveals similar levels of COX III mRNA. Such results indicate that nucleus-encoded 'controller proteins' might regulate the translation of specific mitochondrial mRNAs, or that they might affect the stability of mitochondrial proteins.

Most mitochondrial mRNAs of yeast have a highly conserved dodecamer sequence (5'-AAUAAUAUUCUU-3') at their 3' end. The results of site-directed mutagenesis indicate that a single nuclear gene controls proper post-transcriptional processing at the dodecamer sequence, which is in most cases essential for effective translation.

An alternative approach suggests that expression of some nuclear genes may be regulated by mitochondrial components. Mit^- mutations were induced in yeast strains with an otherwise identical genetic composition. A hybrid-subtraction method was used to clone the DNA which was over-expressed in the mutant yeast cells relative to the normal cells. The resulting differential cDNA library contained sequences corresponding to nuclear ribosomal RNA genes, the 2 μm yeast plasmid, and subunit VI of cytochrome oxidase. The significance of this putative control mechanism is not clear, however. In addition, *petite* mutants lacking any mitochondrial DNA have normal levels of cytochrome c, F_1-ATPase subunits, and ribosomal proteins, all of which are encoded by nuclear genes.

Studies of mitochondrial gene expression in organisms other than yeast are much less well documented. It may be that the expression of mitochondrial genes in mammalian cells is a constitutive, rather than a regulated, process.

8.4.3   How did mitochondria evolve?

It is now widely accepted that mitochondria arose by the symbiotic association of an aerobic prokaryote with a primordial eukaryotic cell. This '*endosymbiont hypothesis*' was presented in figure 1.7. In fact, evidence in favour of such a hypothesis came originally from a comparison of the biochemical features of mitochondria and bacteria (table 8.12). The hypothesis predicts that originally there was no barrier to the transfer of genes from the endosymbiont to the nucleus of the host. With the passage of time, the divergence of the genetic codes used by the two organelles would prevent any further transfer; a few genes would remain isolated in the mitochondrial DNA, which would thus represent an evolutionary dead-end. When one examines the mitochondrial genes of present-day organisms, there is (as expected) no pattern to the retention of genes in the organelle. In a few cases, a copy of a gene has been retained in the mitochondrion as well as transferred to the nucleus. For example, both organelles of *Neurospora* contain a copy of the gene for subunit 9 of the ATP synthase; only the nuclear gene is transcriptionally active, however. The lowest rate

Table 8.12 Comparison of mitochondria and bacteria	Feature	Bacteria	Mitochondria
	Small (70S) ribosomes	Present	Present
	Translation inhibited by chloramphenicol?	Yes	Yes
	Cardiolipin (diphosphatidylglycerol)	Present	Present
	Circular DNA	Present	Present
	Introns in genes	Absent[a]	Present[b]

[a] Absent in *Eubacteria*, but present in certain genes of *Archaebacteria*
[b] Present in some mitochondrial genes of yeast and other lower organisms.

of gene transfer seems to have occurred in higher plants, which have relatively large mitochondrial genomes (table 8.8).

The retention of a separate genetic system in the mitochondrion must represent a metabolic cost to the cell, which has to maintain a complex machinery for replicating, transcribing and translating the mitochondrial genes in a controlled manner. Likewise, the retention of introns in some genes of yeast mitochondria would appear to be energetically wasteful. What advantage might the introns offer that might account for their presence? There is evidence that the introns are, in some cases, non-essential. The common laboratory strain of the yeast, *S. cerevisiae*, has five introns in the gene for cytochrome *b*; in other strains, the gene has only two introns. Furthermore, introns are present in only some mitochondrial genes in fungi, so any advantage would be selectively conferred. In fact, those genes whose introns encode maturases (section 8.3.1) are particularly sensitive to repression by anaerobic growth on glucose. This is exactly what one would expect when the primary transcript of a gene controls its own processing to the mature mRNA. The lack of introns generally in mitochondrial genes suggests that they do not participate in '*exon shuffling*', which would generate chimaeric proteins with novel functions. This process is thought to have been important in the evolution of proteins encoded by nuclear genes (section 10.3).

The mitochondrial genome appears to accumulate nucleotide changes at a rate ten times that of the nuclear genome. This rapid increase in genetic diversity may be attributable to recombination events. Mitochondria contain several (typically, 3–6) copies of their genome, and it is known that the mitochondrial DNA of fungi and plants contains repeated sequences. Such elements would allow homologous recombination to take place between adjacent DNA circles, and thereby bring about an exchange of genes. This explanation would not apply, however, to the genomes of animal mitochondria, which lack repeated sequences.

In conclusion, the endosymbiont hypothesis of the origin and evolution of mitochondria represents a useful working model. Further evidence is needed to substantiate it though.

8.5   Summary

The mitochondrion is the principal site of ATP synthesis within the cell. Its folded inner membrane contains the components of the electron transport chain, which oxidises reduced coenzymes generated by metabolism. In terms of Mitchell's theory of chemiosmosis, electron transport is coupled to the expulsion of protons from the mitochondrial matrix. The resulting electrochemical gradient across the inner membrane drives the synthesis of ATP *via* the F_0F_1-ATPase. The mitochondrion has a complete genetic system, which directs the synthesis of rRNA, tRNA and several of the organelle's proteins; the majority, however, are encoded by nuclear genes. The mitochondrial genetic machinery differs in some remarkable respects from that of the nucleus. Mitochondria import their proteins, synthesised on cytoplasmic ribosomes, by a receptor-mediated mechanism which exploits both ATP and the electrochemical gradient. Cleavable 'address signals' in the mitochondrial proteins ensure their correct location and orientation. New mitochondria probably arise by growth and division of pre-existing organelles. The evolutionary origin of mitochondria can be accounted for by the endosymbiont hypothesis.

8.6   Study questions

1. Which major metabolic pathways are absent in mammalian erythrocytes? (section 8.1.2)
2. How might one use inhibitors to determine whether a mitochondrial protein is encoded by a mitochondrial gene or a nuclear gene? (section 8.3.1)
3. Of what value have mitochondrial mutants been to the study of (a) mitochondrial genetics, (b) mitochondrial protein import? (sections 8.3.2, 8.4.1, and 8.4.2)
4. A genetic mutation causes a complete deficiency of an essential FeS protein of Complex I in certain (human) individuals. What are the likely consequences for intermediary metabolism in the mutant cells? (sections 8.2.1 and 8.1.2)
5. Many cytosolic proteins contain amino acid sequences which potentially could fold to form a positively charged amphiphilic helix. How is it that such proteins are not incorporated into mitochondria? (section 8.4.1)
6. Design an experiment to purify the mitochondrial receptor for apocytochrome c. (section 8.4.1)
7. Brown adipose tissue (BAT) of hibernating mammals has a high rate of thermogenesis (heat production) when stimulated by noradrenaline. BAT mitochondria contain a 32 kDa protein in their inner membrane that is not present in the mitochondria of other tissues. Postulate a mechanism for thermogenesis in BAT. (section 8.2.2)

8.7   Further reading

Books

Whittaker, P. A. and Danks, S. M. (1978) *Mitochondria: Structure, Function and Assembly*, Longman, London.
    (Readable introduction to the subject)
Nicholls, D. G. (1982). *Bioenergetics*, Academic Press, London.
    (Comprehensive account of the chemiosmotic theory)

Reviews

Grivell, L. A. (1983). *Sci. Amer.*, **248** (3), 60–73.
    (Mitochondrial genetics)

Tzagaloff, A. and Myers, A. M. (1986). *Ann. Rev. Biochem.*, **55**, 249–285.
  (Mitochondrial biogenesis)
Hatefi, Y. (1985). *Ann. Rev. Biochem.*, **54**, 1015–1070.
  (Electron transport and oxidative phosphorylation)
Hurt, E. C. and van Loon, A. P. G. M. (1986). *Trends Biochem. Sci.*,
  11, 204–207.
  (Import of mitochondrial proteins)
Attardi, G. (1985). *Int. Rev. Cytol.*, **93**, 93–145.
  (The mitochondrial genome)

9 THE CHLOROPLAST

9.1 Introduction

We have already seen (section 1.2.1) the general features of the ultrastructural organisation of the cells of higher plants. In addition to those organelles shared with animal cell, green plant cells also contain certain structurally and functionally specialised structures: *chloroplasts*, the site of *photosynthesis*; an internal *vacuole* surrounded by its *tonoplast membrane*; and an extracellular *cell wall* composed of cellulose and other polysaccharide material. In this chapter we will say little about the plant vacuole, and nothing about the cell wall. Instead, we will concentrate on the chloroplast, and how its molecular organisation enables it to perform those processes which are vital to all life on Earth: the photosynthetic capture of light energy and fixation of carbon dioxide into carbohydrate, and the associated formation of oxygen.

Chloroplasts are not the only form of plastid found in plants. **Pro-plastids** are immature organelles which give rise not only to chloroplasts but also to *leucoplasts* and *amyloplasts* (plastids with no chlorophyll, little internal membrane, but containing much starch and lipid) and *chromoplasts* (coloured plastids containing high concentrations of pigments called carotenoids). Plants grown in the dark contain immature chloroplasts called *etioplasts*. These organelles lack the stacked internal membranes of mature chloroplasts, but have an extensive and disorganised internal membrane system. Exposure to light switches on numerous genes, in particular those for chlorophyll synthesis, and results in the reorganisation of the internal membranes. However, we shall not give any further consideration to these other plastids.

Recent advances in molecular biology have allowed scientists to characterise many of the protein components that are involved in photosynthesis, as well as the arrangement of the corresponding genetic information in both chloroplast and nucleus. On the basis of such studies, it is now feasible to develop genetically engineered plants which carry out photosynthesis more efficiently, with consequent improvements in agricultural yields. Sunlight is of course the ultimate

Pro-plastids: Precursor organelles with the characteristic double envelope, but lacking chlorophyll and organised internal membranes, and relatively small (spheres of 1 μm diameter, compared with the 5 μm by 2 μm diameter for a mature chloroplast).

energy source, and present-day plants are not particularly efficient in exploiting it: only about 0.2 per cent of the light energy (within the visible spectrum) that falls on the Earth is conserved in a metabolically useful form. It may even be possible to design non-biological systems which mimic the harvesting of light energy that plants achieve. The techniques of molecular biology have also shed new light on the biogenesis of chloroplasts, and on their possible evolutionary origin from a photosynthetic prokaryote that entered into an endosymbiotic relationship with a primordial eukaryotic organism. In the course of evolution, the nucleus has acquired the genes for the synthesis of many chloroplast proteins, and we now have a reasonably clear picture of how these proteins are imported from the cytosol, just as many mitochondrial proteins are (section 8.4.1). Let us start by looking at the detailed structure of the chloroplast.

9.2 Ultrastructure of the chloroplast

9.2.1 The chloroplast contains three distinct membranes

The membrane architecture of a typical chloroplast can be seen in the transmission electron micrograph presented in figure 9.1. A key point is that the organelle is surrounded by a double membrane, composed

Figure 9.1
Ultrastructure of the chloroplast

In this thin section of a mesophyll cell of a maize leaf (*Zea mays*), the double membrane of the chloroplast envelope is apparent. The internal thylakoid membranes are stacked (appressed) to form grana, or are non-appressed (lamellae). The chloroplast stroma (internal matrix) contains several electron-dense granules, and many ribosomes, which are smaller than those of the cytoplasm. × 25 000. [Courtesy of W. W. Thomson and R. M. Leech.]

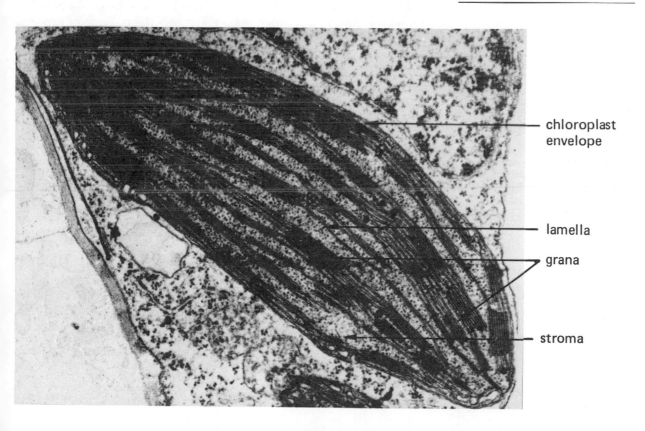

chloroplast envelope

lamella

grana

stroma

of an *external envelope* (or membrane) and an *internal envelope* (or membrane). In this respect, the chloroplast resembles the mitochondrion and the nucleus. In addition, there is an extensive array of internal membranes called *thylakoids*, which form flattened sacs ramifying through the chloroplast. In places there are stacks of thylakoids pressed close together, the *grana*, which may be connected to one another by *lamellae* (non-appressed thylakoids). A diagrammatic representation of this ultrastructural organisation is shown in figure 9.2.

As a consequence of this membrane architecture, the chloroplast contains three distinct spaces: a narrow (10–20 nm) *inter-envelope space*; an extensive **stroma**; and the *intra-thylakoid space* or *lumen*. Since the thylakoids probably originate as outgrowths of the inner envelope that subsequently pinch off into the stroma, the thylakoid lumen is topologically equivalent to the inter-envelope space. We shall see the significance of this topology when we consider the synthesis of ATP by the chloroplast (sections 9.3.4 and 9.3.8).

Stroma: Space between the inner envelope and the thylakoids.

9.2.2   The chloroplast envelopes resemble mitochondrial membranes

One can see numerous striking similarities between the structure and function of the chloroplast envelopes on the one hand, and the mitochondrial membranes on the other. (Compare figure 9.2 with figure 8.3b.) Both organelles are bounded by a smooth outer envelope or membrane that is freely permeable to even quite large solutes. For example, the chloroplast outer envelope allows the passage of proteins up to 10 kDa. This permeability is probably the consequence of

Figure 9.2
Structure of the chloroplast (diagrammatic)

In reality, the stroma would contain many more thylakoids than are depicted here, and the grana would have considerably more appressed thylakoids. (Compare this figure with figures 9.1 and 7.10.)

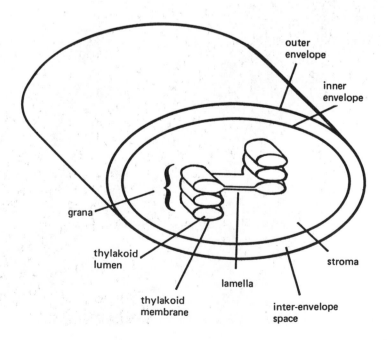

protein-lined pores in the outer envelope similar to those of the mitochondrial outer membrane.

The membrane which represents the permeability barrier between the cytosol and the chloroplast stroma (equivalent to the mitochondrial matrix) is the inner envelope. Unlike the mitochondrial inner membrane, the chloroplast inner envelope is relatively smooth. It does however have a similar very limited permeability to polar solutes. Many of the proteins of the inner envelope are *translocators* responsible for shuttling these solutes between the cytosol and the stroma; some examples are given in sections 9.4.1 and 9.4.2. Water and gases, such as CO_2 and O_2, are able to diffuse freely across the inner envelope.

The thylakoids are the internal membranes of the chloroplast that have the electron transport chain and the proton-driven ATP synthase involved in photosynthetic phosphorylation (*photophosphorylation*). They also contain the pigments involved in the absorption of light energy (*light-harvesting chlorophyll–protein complexes, LHCs*), a system unique to chloroplasts and photosynthetic prokaryotes. The lipid composition of the thylakoid membrane is also distinctive (table 9.1). Phospholipids form a remarkably small proportion of the total, and cholesterol is absent, whereas unique glycolipids, with either one or two galactose residues linked to a diacylglycerol backbone (see structure in figure 9.3) constitute nearly three-quarters of the total. These glycolipids, when dispersed as the pure lipid in aqueous solution, do not form characteristic bilayer structures; clearly, they must interact in some way with the thylakoid membrane proteins to generate membranous arrays. One particularly striking feature of the galactolipids is their very high content of unsaturated fatty acids: up to 90 per cent of the total fatty acid esterified in the galactolipids is constituted by linolenic acid (C18:3). The presence of the three *cis* double bonds (figure 9.3) in this molecule creates a highly fluid environment that may be essential for the proper functioning of the photosynthetic apparatus. Recall that the mitochondrial inner membrane is unusual in having cardiolipin (diphosphatidylglycerol, figure 8.4) as a major lipid constituent. Furthermore, this membrane shares another common feature with that of the thyalkoid: a very low permeability to protons. As we shall see (section 9.3.8), this property is essential to the role of the thylakoid membrane in the maintenance of the proton gradient which is the driving force for photophosphorylation.

9.3 Organisation of the light-harvesting machinery

We will not provide here a detailed account of the '*light reaction*' of photosynthesis; the reader is referred to the general texts quoted in section 9.8. However, we need to consider the biochemical basis of the following features of photosynthesis, and how the molecular components

Table 9.1
Lipid composition of thylakoid membranes

Class of lipid	Moles (per cent)
Monogalactosyl-diacylglycerol	48
Digalactosyl-diacylglycerol	25
Phosphatidyl-glycerol	13
Sulphoquinovosyl-diacylglycerol	8
Phosphatidyl-choline	2
Other phospholipids	4

The proportion of each lipid is given as a molar percentage. Sulphoquinovosyldiacylglycerol is a sulphated glycolipid.

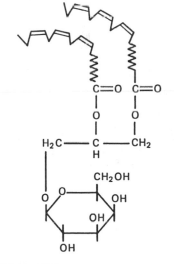

Figure 9.3
Structure of monogalactosyldiacylglycerol

involved are organised in the thylakoid membrane in order to maximise the efficiency of the overall process:

(a)   how the absorption of light energy results in the simultaneous creation of a strong reducing species and a strong oxidising species;

(b)   how the strong reductant passes its electrons to $NADP^+$ in order to generate the NADPH required for the fixation of carbon in the '*dark reaction*';

(c)   how the flow of electrons is coupled to the generation of a proton-motive force, required for the synthesis of ATP that is also essential to the dark reaction;

(d)   how the strong oxidant is able to remove electrons from water and in the process liberate molecular oxygen.

9.3.1   Light-harvesting pigments funnel excitation energy to an unique chlorophyll molecule

The pigments responsible for the absorption of light energy by green plants have been extensively characterised in terms of their structure and spectroscopic properties. The two major groups of pigments are the **chlorophylls** and the **carotenoids**; between them, they can absorb much of the light that falls within the visible spectrum.

Spectroscopic studies have led to the formulation of two important concepts in photosynthesis. Firstly, there are two separate light-absorbing *photosystems* involved, one activated maximally by light of wavelength 700 nm and one by shorter wavelength light (680 nm or below). Secondly, a discrete photosynthetic unit is not associated with a single molecule of chlorophyll, but with a network of 300 or so chlorophylls. Absorption of a photon by any of these *antenna pigments* results in promotion of an electron to a higher orbital and the consequent formation of an excited state. The excitation energy is transferred very rapidly from one to another of the chlorophyll molecules that comprise the light-harvesting complex, within a time-span of picoseconds (10^{-12}s). Eventually it reaches a chlorophyll *a* molecule (or a special pair of molecules) in an unique environment, referred to as *P*700 (for Photosystem I) or *P*680 (for Photosystem II). The light-harvesting complexes (LHCs) contain not only the light-absorbing pigments, but also specific polypeptides. In some as yet uncharacterised manner, the LHC polypeptides in the thylakoid membrane orientate the pigments so as to facilitate the energy transfer to the reaction-centre chlorophyll. The LHC of Photosystem II has been partially characterised: it contains 15 or so apoproteins (polypeptides) which bind non-covalently to the antenna pigments, and form hexameric complexes in the thylakoid membrane. The LHC polypeptides share a high degree of amino acid sequence homology, and probably are coded for by a multi-gene family in the nuclear DNA.

Chlorophyll: Haem-like compound complexed to magnesium ion.
Carotenoid: Isoprenoid lipid with an extended system of alternating double bonds.

*P*700 (*P*680):   Chlorophyll pigment which absorbs light maximally at *700* nm (or *680* nm).

9.3.2   Photosystem I transfers electrons to NAD$^+$

The unique chlorophyll molecule ($P700$) at the reaction centre of Photosystem I is thus converted to an excited state ($P700^*$). The electron involved, promoted now to a higher orbital, is then transferred to a series of electron acceptors which can be alternately reduced and oxidised. The identities of some of these components are currently (1988) unclear, though they probably include a special chlorophyll *a*, a quinone derivative, and several iron–sulphur protein complexes. Whatever their nature, we do know that the electrons finally arrive at **ferredoxin.** Reduced ferredoxin can transfer its electrons to NADP$^+$ in a reaction catalysed by a specific flavoprotein, ferredoxin–NADP$^+$ reductase.

> **Ferredoxin**: Extrinsic protein on the stromal side of the thylakoid membrane; contains two iron atoms and two labile sulphur atoms per molecule, and acts as a one-electron carrier.

One consequence of this sequence of events is to leave the reaction-centre chlorophyll deficient in one electron ($P700^+$). Neutrality is restored by capture of an electron, but from where?

9.3.3   Photosystem II donates electrons to Photosystem I

The unique chlorophyll ($P680$) at the reaction centre of Photosystem II absorbs light energy as previously described, and the promoted electron of the excited state ($P680^*$) is donated ultimately to Photosystem I. The initial electron acceptor is probably **pheophytin**, followed by two distinct quinone derivatives (Q_A and Q_B). The electrons are then channelled into a transmembrane complex of carriers that closely resembles Complex III of mitochondria (section 8.2.1); the chloroplast components are *plastoquinone*, a specific iron–sulphur protein (Rieske protein), cytochrome b_6, and cytochrome *f* (an analogue of mitochondrial cytochrome c_1). The chloroplast complex furnishes pairs of electrons, taken from the preceding one-electron carriers, to the soluble copper-containing protein *plastocyanin*, at the luminal face to the thylakoid membrane. Reduced plastocyanin then passes its electrons to $P700^+$.

> **Pheophytin**: Chlorophyll derivative which lacks Mg^{2+}.

9.3.4   The two photosystems are functionally interconnected

We have seen how Photosystem II donates electrons to Photosystem I *via* an electron transport chain in the thylakoid membrane. The usual representation of this functional relationship is the classical *Z-scheme*, depicted in figure 9.4. Non-cyclic electron flow in this system results in the light-activated extraction of electrons from water (to form O_2, section 9.3.9), with the eventual reduction of NADP$^+$ to NADPH and the generation of ATP:

$$NADP^+ + H_2O + ADP + P_i$$
$$\downarrow$$
$$NADPH + H^+ + ATP + \tfrac{1}{2}O_2$$

The stoichiometry of the process is such that 2 NADPH and 2 or 3 ATP are synthesised for each pair of electrons activated by Photosystem I and Photosystem II.

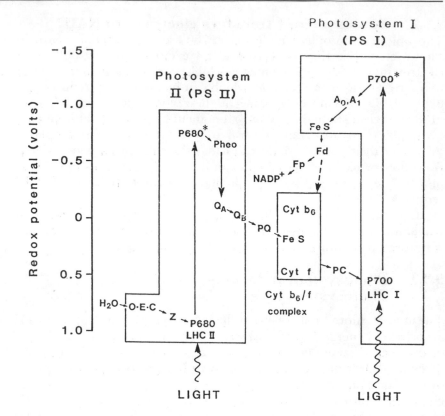

An alternative pathway for the transfer of electrons from activated $P700$ is shown by the dashed line in figure 9.4. This cyclic electron flow proceeds from reduced ferredoxin to the cytochrome b/f complex, and thence back to $P700$. The consequence is that NADPH is not formed, but electron transport does contribute to the establishment of the proton gradient, so that ATP can be generated. This electron flow around Photosystem I may permit the maintenance of an optimal level of ATP relative to that of NADPH, since both are required for the fixation of carbon dioxide in the 'dark reaction' of photosynthesis, but at different stoichiometries (3 ATP and 2 NADPH per CO_2 reduced).

9.3.5 The Z-scheme components form three membrane-spanning complexes

The protein constituents of Photosystem I, Photosystem II, and the cytochrome b/f complex are oriented in the thylakoid membrane so as to optimise electron flow and the acidification of the luminal space. We now have considerable information on the identity of the proteins involved and their topological distribution within the membrane, information derived from several experimental approaches:

(a) Preparation of the individual complexes by differential extraction of thylakoid membranes.

(b) Characterisation of their polypeptide composition by SDS-PAGE and Western blotting.

(c) Purification of some of the proteins to homogeneity, followed by amino acid sequencing.

(d) Cloning of several of the genes concerned, followed by determination of their base sequences.

(e) Computer predictions (*hydropathicity plots*) as to which amino acid sequences might form membrane-spanning regions.

(f) Freeze-fracture studies by electron microscopy.

(g) Formation of right-side-out and inside-out vesicles from thylakoid membranes (figure 9.5). The orientation of the membrane components of such vesicles can be determined by the use of non-penetrating probes (antibodies and proteases, in particular).

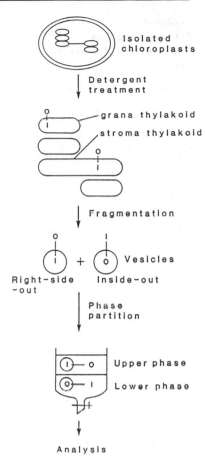

The picture that emerges is that of an asymmetric distribution of protein constituents across the thylakoid membrane, in a manner reminiscent of the mitochondrial electron transport chain (figure 8.6). Further analogies have been drawn between Photosystem II and the reaction centre of certain photosynthetic bacteria, which has been crystallised and its three-dimensional structure determined by X-ray diffraction. A simplified representation of the structure of the reaction centre of Photosystem II is shown in figure 9.6. Each of the three major proteins (D1, D2 and cytochrome b_{559}) spans the membrane; in fact, computer analysis of the amino acid sequence of D1 suggests that it might contain seven transmembrane helical regions. Certain classes of herbicide (such as DCMU, dichlorophenyldimethylurea) bind to D1 at the stromal face of the thylakoid membrane and block electron transfer to the associated two-electron quinone acceptor, Q_B. We shall no doubt soon have a similarly detailed picture of the organisation of Photosystem I and the cytochrome b/f complex.

Figure 9.5
Preparation of vesicles from thylakoid membranes

On disruption, the fragments of thylakoid membrane re-seal to form two populations of vesicles. The non-appressed membranes of the stromal thylakoids (lamellae) form mainly right-side-out vesicles; these partition into the upper phase of a two-phase aqueous polymer system. The appressed membranes of the grana thylakoids form inside-out vesicles, which partition into the lower phase. A transmembrane component (O–I) is shown with some of its structure originally exposed to the stroma (O–) or to the lumen (–I).

9.3.6 Photosystem I and Photosystem II are asymmetrically distributed along the thylakoid membrane

When the techniques became available for the preparation of thylakoid membrane vesicles (figure 9.5), it was rapidly confirmed what was previously suspected from freeze-fracture EM analysis: that the two photosystems were not evenly distributed along the thylakoid membrane. Photosystem II (PSII) was enriched in the appressed thylakoids present in grana, whereas Photosystem I (PSI) predominated in the lamellae (non-appressed thylakoids). This spatial separation seems rather curious when one considers that the two photosystems have to interact, with PSII donating electrons to PSI. Presumably plastoquinone can diffuse rapidly in the membrane to transfer electrons from PSII to the cytochrome b/f complex, and from there plastocyanin can transfer the electrons to PSI by rapid diffusion along the luminal face of the thylakoid membrane. But what can be the functional significance behind this lateral separation of the two photosystems?

Since the two systems operate in series, it is important that under

Figure 9.6
Schematic diagram of
Photosystem II reaction
centre

The direction of electron transport is
shown by arrows. The components
are abbreviated as follows:
D1, 32 kDa protein which binds a
ubiquinone-like compound (Q_B);
D2, 34 kDa protein which binds a
menaquinone-like compound (Q_A);
Cyt b_{559} (hatched), cytochrome b_{559};
Chl, chlorophyll; Pheo, pheophytin;
Fe, uncharacterised non-haem-iron
component; PQ, plastoquinone. (For
the sake of simplicity, numerous other
components have been omitted.)

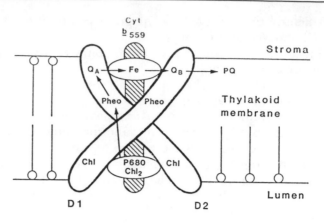

most circumstances, the chloroplast ensures an even distribution of light
energy between PSI and PSII. When PSII is active, the high levels of
reduced plastoquinone activate a membrane-bound protein kinase
which phosphorylates the light-harvesting complex (LHC) polypeptides
on their exposed amino-terminal domains. The phosphorylated LHCs
migrate in the plane of the membrane, away from the PSII reaction
centres and towards the PSI reaction centres, and thus tip the
absorption of light energy in favour of PSI (the so-called 'spillover'

Figure 9.7
Migration of light-harvesting
complexes between
Photosystems I and II

A membrane-bound kinase, activated
by reduced plastoquinone, adds
phosphate groups on to serine and/or
threonine residues of the polypeptides
of the light-harvesting
chlorophyll–protein complexes
(LHCs). The LHCs (filled triangles)
migrate from PSII (hatched circles)
in the grana to PSI (open circles) in
the lamellae. The process is reversible.
(For simplicity, the complexes are
shown in only two thylakoid
membranes.)

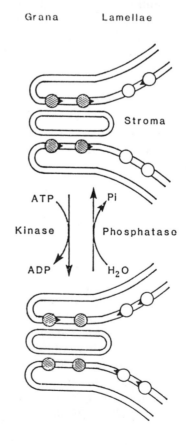

effect). Conversely, when the LHCs are dephosphorylated by a phosphatase enzyme, they migrate back to the stacked regions of the thylakoid, where they associate with the PSII reaction centres. These interactions are summarised in figure 9.7.

9.3.7 Electron transport generates a proton gradient

We saw earlier (section 8.2.2) how the passage of electrons along the mitochondrial electron transport chain was coupled to the expulsion of protons from the organelle's matrix to the inter-membrane space; the resulting pH gradient and charge separation represented a source of free energy that could drive ATP synthesis (oxidative phosphorylation). A similar mechanism underlies photosynthetic phosphorylation. Here, however, the protons are translocated from the stroma across the thylakoid membrane, rather than to the inter-envelope space.

Experimental evidence for such a mechanism came from the classical studies of **Jagendorf** in 1969. A representative experiment is depicted in figure 9.8. Jagendorf was able to demonstrate that under these conditions, ATP was synthesised by the chloroplasts, even in the absence of light. Such experimental evidence connected with photophosphorylation was some of the first to support Mitchell's chemiosmotic hypothesis, some years before any convincing parallel studies were made on mitochondrial oxidative phosphorylation.

What is the source of the protons translocated across the thylakoid

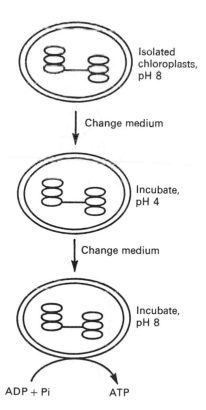

Isolated chloroplasts, pH 8

Change medium

Incubate, pH 4

Change medium

Incubate, pH 8

ADP + Pi   ATP

**Figure 9.8
ATP synthesis by isolated chloroplasts**

Isolated chloroplasts were incubated for several hours in the dark in an acidic buffer (pH 4). When the medium was changed to a slightly alkaline buffer (pH 8), the chloroplasts catalysed the synthesis of ATP from ADP and P_i present in the medium, in a reaction driven by the proton gradient across the thylakoid membrane.

membrane? Two protons are generated in the thylakoid lumen by the oxidation of a molecule of water (section 9.3.9). Additional protons are taken up at the stromal face of the thylakoid membrane and/or given up at the luminal face in the following steps: the reduction of plastoquinone; the oxidation of reduced plastoquinone by the cytochrome b/f complex; and the reduction of $NADP^+$. The result is a gradient of 3–4 pH units across the thylakoid membrane. In chloroplasts, the transfer of positive charge across the membrane is compensated for by an exchange with Mg^{2+} ions or by the simultaneous uptake of Cl^- ions. The result is that the charge separation is negligible, whereas in mitochondria both the pH gradient and the charge separation make a significant contribution to the free energy of the proton gradient.

9.3.8   The proton gradient across the thylakoid membrane drives ATP synthesis

Freeze-fracture EM studies of the thylakoid membrane revealed numerous knob-like projections on the stromal surface. By analogy with the mitochondrial F_0F_1-ATPase (figure 8.3c), these were taken to be the chloroplast equivalent, **CF_0CF_1-ATPase**. Estimates of CF_1 in thylakoid membrane vesicles (figure 9.5) suggest that three-quarters of the ATP synthase complexes are located primarily in the lamellae; this lateral asymmetry resembles the distribution of Photosystem I.

CF_0CF_1-ATPase: Thylakoid-associated *coupling factor*, otherwise called chloroplast *ATP synthase*.

Figure 9.9
Structure of the chloroplast ATP synthase (CF_0CF_1)

The CF_1 complex contains five different subunits (molecular masses in parentheses): α (58 kDa), β (57 kDa), γ (35 kDa), δ (21 kDa), and ε (35 kDa). The CF_0 complex contains one copy each of subunits I (19 kDa), II (15 kDa), and IV (19 kDa), and six copies of subunit III (8 kDa) which form a transmembrane pore. Passage of protons through the CF_0CF_1 complex is coupled to the synthesis of ATP. (Compare this structure with that of the mitochondrial F_0F_1-ATPase, figure 8.12.)

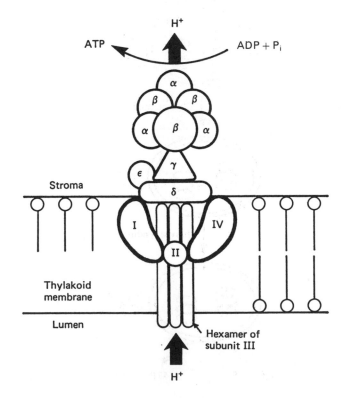

CF$_1$, the multi-subunit complex which projects from the thylakoid membrane, can be separated from the integral membrane complex CF$_0$, with which it is associated. A model of the organisation of the chloroplast ATP synthase is presented in figure 9.9.

Reconstitution studies and inhibition studies with photoactivatable analogues of ATP suggest that it is the β subunits (and possibly also the α subunits) that are catalytically active. Inhibitors of proton flow, such as DCCD (dicyclohexylcarbodiimide), bind tightly to the 8 kDa proteolipid of CF$_0$ that forms a hexameric complex spanning the thylakoid membrane and connected to CF$_1$. The genes for several of the polypeptides of the chloroplast ATP synthase have been cloned (table 9.2), and mutant plants genetically deficient in one or other subunit have been characterised. However, we still do not have a clear picture of the functional interactions between the various subunits during proton-driven photophosphorylation. Also, the precise stoichiometry of the overall process is still disputed: the amount of ATP synthesised per pair of electrons transported has been variously reported as 1.0, 1.3, and 2.0 molecules.

9.3.9 Photosystem II is responsible for the formation of oxygen from water

Excitation by light of $P680$ and the subsequent transfer of the excited electron to pheophytin results in an electron-deficient reaction centre, $P680^+$. This species is a strong enough oxidant to extract electrons from

Complex	Gene	Identity of gene product
Photosystem I	psaA	P700-Chl a-binding protein
	(—)	(17 kDa and 11 kDa polypeptides)
Photosystem II	psbA	D1 (32 kDa quinone-binding protein)
	psbB	51 kDa Chl a-binding protein
	psbC	44 kDa Chl a-binding protein
	psbD	D2 (34 kDa quinone-binding protein)
	psbE	Cytochrome b_{559}
Cytochrome	petA	Cytochrome f
b/f complex	petB	Cytochrome b_6 (b_{563})
	petD	15 kDa polypeptide
ATP synthase	atpA	α subunit of CF$_1$
(CF$_0$CF$_1$)	atpB	β subunit of CF$_1$
	atpE	δ subunit of CF$_1$
	atpF	Subunit I of CF$_0$
	atpH	Subunit III of CF$_0$

**Table 9.2
Chloroplast genes that encode photosynthesis-related proteins**

The genes have been allocated a three-letter code related to their function (for example, *pet* = photosynthetic electron transport). The arrangement of genes on the circular chloroplast DNA is shown in figure 9.14. Chl = chlorophyll.

water (*via* an uncharacterised donor, Z, and the oxygen-evolving complex), and thereby generate molecular oxygen:

$$2H_2O + 4P680^+ \rightarrow 4P680 + 4H^+ + O_2$$

How are the four electrons involved extracted by the oxygen-evolving complex and then transferred to the reaction centre of PSII? Inside-out vesicles prepared from grana-derived thylakoid membranes (figure 9.5) have provided a useful experimental system. Extrinsic proteins can be removed by selective washing procedures, and oxygen-evolving capacity reconstituted by adding back the purified proteins; by such means, proteins of 33 kDa, 23 kDa, and 16 kDa have been characterised. Their possible association with one another and with the PSII reaction centre is shown in figure 9.10. Studies of oxygen evolution by dark-adapted chloroplasts suggest that the oxygen-evolving complex may pass through five successive oxidation states of increasing positive charge (*S states*) before the four electrons are acquired from two molecules of water. However, the precise mechanism whereby this process is achieved is unclear.

9.4   Organisation of carbon-fixation metabolism

The net result of the '*light reaction*' of photosynthesis is to generate NADPH and ATP in the stroma of the chloroplast. How are these compounds used to fix carbon, by reducing CO_2 to the level of carbohydrate? We do not have space here for a detailed consideration of the metabolic pathways involved. These were elucidated by the pioneering studies of **Calvin** and **Bassham,** who followed the incorporation of radioactive CO_2 into labelled photosynthetic products (triose phosphates) which were characterised by two-dimensional paper chromatography. A summary of the *Calvin cycle* reactions, which occur in the stroma of the chloroplast, is presented in figure 9.11.

Figure 9.10

The oxygen-evolving complex of Photosystem II (PSII)

Two molecules of water are oxidised to a molecule of oxygen, four protons ($4H^+$) and four electrons ($4e^-$). The 33 kDa protein probably has two more manganese (Mn) ions associated with it. The structure of PSII is depicted in figure 9.6. Each light-harvesting chlorophyll–polypeptide complex (LHC) contains about 25 molecules of apoprotein and about 150 molecules of chlorophyll (*a* and *b*).

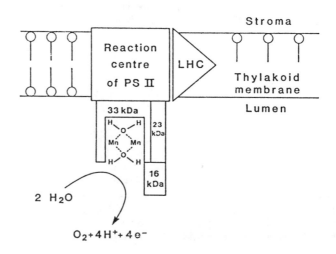

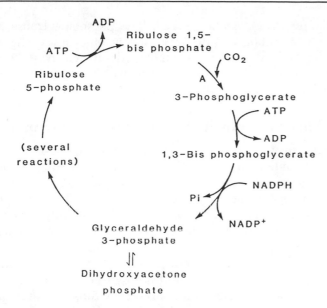

Figure 9.11
The Calvin cycle (simplified)

In the photosynthetic fixation of CO_2, the immediate product is 3-phosphoglycerate (two molecules), which is reduced to glyceraldehyde 3-phosphate. For each molecule of CO_2 fixed, three molecules of ATP and two of NADPH are required. The regeneration of ribulose 1,5-bisphosphate (CO_2 acceptor) involves a series of reactions that resembles the pentose phosphate pathway. Each reaction shown here is catalysed by a specific enzyme; the rate-limiting step (A) of the Calvin cycle is that catalysed by ribulose bisphosphate carboxylase.

9.4.1 Triose phosphates are exported to the cytosol

Glyceraldehyde 3-phosphate and dihydroxyacetone phosphate are exported from the stroma of the chloroplast to the cytosol of the photosynthetic cell. One consequence of this transport is that phosphate would be continually lost from the organelle, if it were not replaced. In face, the inner envelope of the chloroplast membrane has an integral protein which translocates triose phosphate in exchange for inorganic phosphate (figure 9.12). The significance of this *phosphate translocator* is that it enables NADPH and ATP generated by photosynthetic electron transport to be delivered to the cytosol. NADPH itself cannot cross the chloroplast membrane, and ATP as such is probably not exported

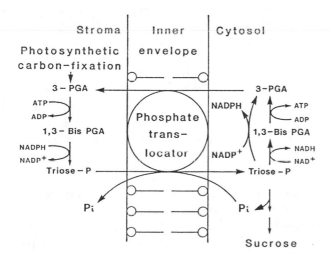

Figure 9.12
Metabolic role of the phosphate translocator

In chloroplasts undergoing photosynthesis, glyceraldehyde 3-phosphate dihydroxyacetone phosphate (Triose-P) formed in the stroma are exchanged for inorganic phosphate (P_i). Triose-P in the cytosol can give rise to ATP, NADH, and NADPH in reactions involving 1,3-bisphosphoglycerate (1,3-BisPGA) and 3-phosphoglycerate (3-PGA). P_i in the stroma is used to regenerate ATP from ADP by photophosphorylation.

during photosynthesis (section 9.4.2). The phosphate translocator also transports 3-phosphoglycerate back into the chloroplast (figure 9.12). The net effect of this shuttle is to export reducing equivalents and ATP from the stroma to the cytosol. (Compare the glycerol 3-phosphate shuttle of mitochondria; figure 1.6.)

9.4.2   The inner envelope contains other translocators

In addition to the phosphate translocator, the chloroplast inner envelope also contains an *ADP–ATP translocator*. This integral membrane protein, analogous to that of mitochondria (table 8.2), is not thought to contribute significantly to the export of ATP formed in the chloroplast during photosynthesis. Instead, it probably acts to import ATP into the organelle from the cytosol, the ATP being generated by glycolysis and respiration during the hours of darkness.

Also present is a membrane-bound *glucose transporter*. Some of the triose phosphate formed during photosynthetic carbon-fixation is converted by chloroplast enzymes into starch. Subsequent degradation of this polysaccharide by hydrolysis results in glucose and maltose, which are exported to the cytosol via the glucose transporter.

The chloroplast inner envelope contains at least three other translocators required for photorespiration (figure 7.9): (1) a *glycolate transporter* (to deliver glycolate to the cytosol); (2) a *glycerate transporter* (to take up glycerate from the cytosol); and (3) a *dicarboxylate transporter* (to exchange 2-oxoglutarate and glutamate in order to refix the NH_3 formed during the oxidation of glycine).

9.4.3   Other metabolic reactions are localised within chloroplasts

Chloroplasts make a significant contribution to the metabolism of plant cells, over and above their role in the 'light reaction' and the 'dark reaction'. In summary, they are responsible for:

(a)   conversion of NO^{2-} to NH_3 (nitrogen assimilation) by reduced ferredoxin in the stroma;

(b)   conversion of SO_4^{2-} to the sulphur-containing amino acid, cysteine;

(c)   synthesis of long-chain fatty acids (palmitic, C16:0; stearic, C18:0; oleic, C18:1) in the stroma;

(d)   inactivation of superoxide by superoxide dismutase in the stroma;

(e)   inactivation of H_2O_2 by ascorbic acid (vitamin C);

(f)   synthesis of galactolipids (section 9.2.2) in the envelope membranes.

Several species of plants have evolved specialised metabolic pathways which serve to enhance the supply of CO_2 to the rate-limiting enzyme in carbon-fixation in the chloroplast, ribulose 1,5-bisphosphate carboxylase. In some plants (**C₄ species**), phospho*enol*pyruvate is

C₄ species: Certain tropical plants, such as sugar cane and maize, which actively synthesise the four-carbon compound malate by carboxylation.

converted (*via* oxaloacetate) to malate in the mesophyll cells. The malate is delivered to the bundle sheath cells, where decarboxylation to pyruvate generates extra CO_2 for the Calvin cycle. In the *Crassulaceae* and certain other plant families, a similar metabolic sequence operates; however, here the malate is accumulated in the vacuoles of all photosynthetic cells at night, and is only decarboxylated to liberate extra CO_2 for the chloroplast during the daytime.

9.5 Biogenesis and evolution of chloroplasts

In the course of sexual reproduction and development in plants, a single cell (the zygote) ultimately gives rise to all the tissues of the mature organism, and hence to all its organelles. As far as we know, chloroplasts do not originate *de novo*, in the way that lysosomes do, for example. Instead, there is a mechanism for incorporation of new material (lipids and proteins) into immature chloroplasts, and subsequent division of the mature organelles. Furthermore, the evolutionary origin of chloroplasts is of some importance. Did they evolve from *cyanobacteria* (blue-green algae) that became associated with a primordial eukaryote in an endosymbiotic relationship, as depicted in figure 1.7? Here, we consider the biogenesis and evolutionary lineage of the chloroplasts of higher plants.

9.5.1 Cyanobacteria-like organelles ('cyanelles') resemble primitive chloroplasts

Certain photosynthetic, unicellular organisms (including *Cyanophora paradoxa*) are eukaryotic cells that contain inclusions which resemble cyanobacteria. The evidence for such an identification is convincing: there is a thin peptidoglycan cell wall; there are whorls of unstacked photosynthetic membranes; and the surface of these membranes is associated with **phycobilisomes**. However, the inclusions cannot be cultured *in vitro*, and their content of DNA is only 10 per cent of that of a typical cyanobacterium. For these reasons, it has been concluded that such inclusions originated from the development of an endosymbiotic relationship between a cyanobacterium and an early eukaryote; they have been given the name *cyanelles*.

The functional parallels between cyanelles and chloroplasts are striking. They both contain chlorophyll, carotenoids, Photosystem I (with *P*700), cytochromes *b* (b_{559}) and *f*, ferredoxin, ATP synthase, and ribulose 1,5-bisphosphate carboxylase. However, cyanelles lack an identifiable Photosystem II and plastocyanin, but have in addition a soluble cytochrome *c* (c_{553}), which probably functions as a mobile electron carrier in a manner analogous to that of plastocyanin in higher plants. Cyanelles fix CO_2 photosynthetically, and export glucose, maltose and sucrose to the surrounding cytosol. In fact, *Cyanophora paradoxa* is absolutely dependent on this source of metabolic energy.

The relatively small cyanelle genome (mass about 1.2×10^5 kDa)

Phycobilisomes: Accessory light harvesting components characteristic of cyanobacteria.

is sufficient to code for only 100 or so average-length polypeptides. This loss of genetic potential compared with that of free-living cyanobacteria is presumably a reflection of a symbiotic relationship, whereby the host cell provides many of the needs (structural and metabolic) of the organelle. The synthesis of RNA in cyanelles is inhibited by rifampicin, just as it is in chloroplasts and prokaryotic organisms, whereas protein synthesis in the host cell is blocked by cycloheximide, as in eukaryotic cells.

So the cyanelle appears to resemble a chloroplast which has developed from a prokaryotic precursor. However it is unlikely that such an evolutionary origin occurred only once. The red algae probably evolved thus from an association with a cyanobacterium, whereas the chloroplasts of land plants may have derived from another prokaryote (for example, *Prochloron*) with chlorophyll *a* and *b* and with stacked thylakoids.

9.5.2   Chloroplasts arise from pre-existing organelles

In developing plant cells, immature forms of chloroplasts and other plastids are present in the form of *pro-plastids* (section 9.1). In photosynthetic cells, chlorophyll synthesis is activated by light in a process controlled by photoreceptors such as *phytochrome*. Thylakoids are thought to originate by invagination of the inner envelope to form flattened vesicles, which then pinch off into the stroma. The association of adjacent thylakoids to form grana may be mediated by interactions between the light-harvesting complex polypeptides, which project about 20 amino acids of their *N*-terminal portions out into the stroma.

Mature chloroplasts add new lipid and protein to their existing structure. They are able to synthesise many lipids *de novo*, including fatty acids and the galactolipids (section 9.4.3). However, they are unable to synthesise linolenic acid (C18:3) or to introduce multiple double bonds into glycerolipids, a process which is carried out by enzymes of the endoplasmic reticulum. The chloroplast genome can potentially code for over 100 polypeptides, but the majority of the organelle's proteins are encoded by nuclear genes, synthesised on cytoplasmic ribosomes, and post-translationally imported into the organelle (section 9.5.3). By such means the chloroplast enlarges its structure until it is able to undergo division by a form of binary fission (figure 9.13). However, we still do not understand the detailed structure of the division apparatus and the mechanism (if any) for the partitioning of chloroplast DNA to each of the daughter organelles.

9.5.3   The chloroplast genome codes for ribosomal components and photosynthesis-related proteins

The chloroplast DNA exists as multiple copies of a circular double-stranded molecule of (on average) about 150 kb, sufficient to code for about 120 average-size polypeptides, of which more than 30 have been identified. It differs from the nuclear genetic material in having a large segment of inverted repeat sequence, and in lacking histones bound to the DNA. The gene products of the organelle's genome include transfer

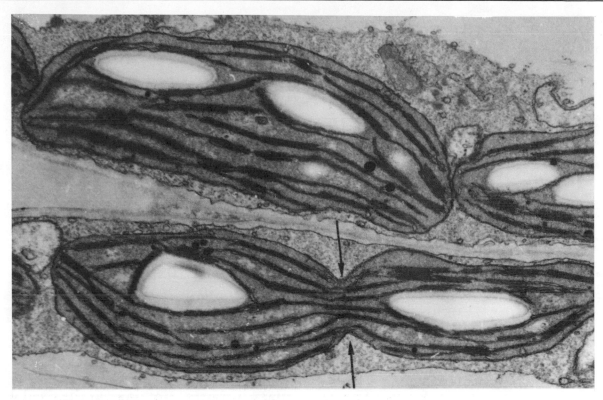

Figure 9.13
Chloroplast undergoing division

In this thin section of a leaf from a 7-day old wheat plant, the lower cell is seen to contain a mature chloroplast undergoing division (arrows). × 20 000. [Original electron micrograph by K. A. Phalt-Aloia; courtesy of R. M. Leech.]

RNAs, ribosomal RNAs (23S and 16S), ribosomal proteins (at least 10 polypeptides of the small subunit and at least 7 polypeptides of the large subunit), initiation and elongation factors, RNA polymerase (3 subunits), ribulose bisphosphate carboxylase (large subunit), and several proteins associated with photosynthesis (table 9.2).

The location of most of these gene loci is known; a simplified gene map is shown in figure 9.14. Note that structural genes are present on both strands of the chloroplast DNA; intervening sequences (introns) within such genes are present, but are relatively rare. The chloroplast genes are transcribed into poly(A)− mRNA species, which are translated by ribosomes within the organelle into the mature-size polypeptides. (Contrast the chloroplast proteins encoded by nuclear genes: they are translated by free cytoplasmic ribosomes from poly(A)+ mRNA into larger precursors with N-terminal extensions; section 9.5.4). Northern blotting has demonstrated that some of the adjacent genes are co-transcribed into polycistronic mRNA: for example, a major spliced transcript of 3.3 kb directs the synthesis of CF_0 subunit III, CF_0 subunit I and the α subunit of CF_1 (genes *atp*H, *atp*F, and *atp*A respectively). Unlike mitochondria, chloroplasts use the 'universal' genetic code, with AUG serving as the initiation codon.

The cloning of the chloroplast genes mentioned above means that they are now amenable to site-directed mutagenesis in order to generate modified proteins with enhanced properties. Particular attention is

Figure 9.14
Map of the chloroplast genome

The locations of the gene loci indicated are those for pea (*Pisum sativa*); other species differ in the relative positions of the genes. The abbreviations used are those in table 9.2; *rbcL,* large subunit of ribulose bisphosphate carboxylase. The direction of transcription is indicated by arrows. (Note that the map is not complete. Protein-coding sequences lie within the unlabelled regions, but the identity of the gene products is not known at present (1988). However, the expression of cloned DNA fragments will allow researchers in the near future to identify the chloroplast proteins encoded by these unassigned reading frames.)

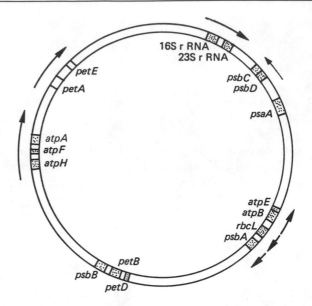

being focused on improving the efficiency of ribulose bisphosphate carboxylase, the rate-limiting enzyme in the carbon-fixation stage of photosynthesis (section 7.2.3).

9.5.4 The chloroplast imports proteins synthesised on cytoplasmic ribosomes

The implication of section 9.5.3 is that many of the proteins involved in photosynthesis in the chloroplast are encoded by nuclear genes: for example, plastocyanin and ferredoxin, the γ and ε subunits of CF_1, subunit II of CF_0, and the small subunit of ribulose bisphosphate carboxylase. In these cases, the proteins released from cytoplasmic ribosomes must somehow find their way to their appropriate location within the organelle. What is the nature of the 'address system' involved, how do chloroplast proteins avoid being taken up by other organelles, and how do chloroplasts avoid taking up proteins intended for other subcellular destinations?

An important clue to the mechanisms involved came from cell-free protein synthesis studies using a wheat-germ extract system and poly(A) + (messenger) RNA from the photosynthetic alga, *Chlamydomonas reinhardtii*. This system synthesised the small subunit of ribulose bisphosphate carboxylase as a precursor that was 3.5 kDa larger than the mature protein (14 kDa). This extension was much larger than the typical signal sequence of proteins synthesised by membrane-bound ribosomes (section 4.2.1). When the translation products were added to isolated chloroplasts, the small subunit was taken up, processed to the mature form, and integrated with the chloroplast-synthesised large subunit into the catalytically active *holoenzyme*. Such studies demonstrated convincingly that the import of chloroplast proteins from the cytoplasm was a post-translational event, in marked contrast to the co-translational passage of secreted proteins across the membrane of the rough endoplasmic reticulum.

The general features of the chloroplast system resemble those of the mitochondrial system (section 8.4.1). (There is an added complication though: some chloroplast proteins, such as plastocyanin in the thylakoid lumen, have to cross *three* membranes before they reach their final destination.) In all cases studied, the precursor polypeptides have an *N*-terminal extension that ranges in size from 34 to 66 amino acids; such extensions are generally referred to as '*transit sequences*'. A common feature of these chloroplast–protein sequences is their relatively high content of positively-charged amino acids (arginine and lysine) and hydroxy amino acids (serine and threonine), with the consequence that the sequence overall has hydrophilic character. (Contrast the hydrophobic nature of the signal sequences of secreted proteins, section 4.2.1.)

The positively charged transit sequences associate ionically with the highly negatively charged chloroplast outer envelope. However, the latter also contains specific receptors for chloroplast proteins, as might be expected. Inward transport of bound polypeptides, which may occur at sites of adhesion between the inner and outer envelopes, is dependent upon a supply of ATP; no electrochemical energy is involved, in the way that it is in the mitochondrial import system (section 8.4.1). The transit peptidases that remove the *N*-terminal extensions have not yet been extensively characterised. They seem to be soluble, metal-containing endoproteases which generate intermediates in a processing mechanism that involves at least two steps. There is also evidence for two such enzymes, one localised in the stroma and one in the lumen of the thylakoid.

9.6 Summary

Chloroplasts are the photosynthetic organelles of plants. They are bounded by a double membrane; within their matrix, or stroma, lie thylakoid membranes, often stacked to form grana. Light energy is absorbed by chlorophyll and other pigments present in transmembrane light-harvesting complexes consisting of polypeptides and pigments. The excitation energy is passed rapidly to an unique chlorophyll molecule at the reaction centre of either Photosystem II (*P*680) or Photosystem I (*P*700); these two are coupled by the cytochrome b/f complex. Electron transport through these transmembrane complexes results in the accumulation of protons in the thylakoid lumen. The proton gradient across the thylakoid membrane drives the synthesis of ATP by the CF_0CF_1 ATP synthase (photophosphorylation). Photosystem I also reduces $NADP^+$, whereas Photosystem II oxidises H_2O to molecular oxygen. The ATP and NADPH generated during the 'light reaction' of photosynthesis are used by the chloroplast to fix CO_2 into carbohydrate *via* the Calvin cycle of reactions in the stroma. The products of this 'dark reaction', mainly triose phosphates, are exported to the cytosol, where they are used to generate ATP, reduced coenzymes, and more complex carbohydrates, such as sucrose. Chloroplasts are thought to have evolved from cyanobacteria-like prokaryotes. They arise either from the differentiation of immature precursors called pro-plastids, or by the division of mature chloroplasts. Their maturation involves the incorporation of endogenously synthesised lipids and proteins, as well as the import of cytoplasmically synthesised proteins encoded by nuclear genes. Numerous chloroplast genes have been cloned and mapped on the organelle's circular genome.

9.7   Study questions

1. One gram of a typical leaf contains about 300 million chloroplasts with a total volume of about 30 μl. The chloroplasts occupy approximately 20 per cent of the cell volume, which is on average 10^4 μm^3. How many chloroplasts does a typical leaf cell contain? (1 l is equivalent to 1 dm^3.)

2. Justify the assertion (section 9.5.3) that the chloroplast DNA (150 kb) can code for 120 polypeptides (average mass 30 kDa). (The average mass of an amino acid in a polypeptide is 0.12 kDa.)

3. The enzyme that catalyses the first step in CO_2 fixation, ribulose bisphosphate carboxylase, can also act as an oxygenase. What would be the disadvantage of the latter activity to the chloroplast? (sections 9.4.1 and 7.2.3)

4. Name the 5 transmembrane complexes present in the thylakoid membrane. Describe the movement (if any) of electrons and protons within each complex. (figures 9.4, 9.6 and 9.10)

5. Purple bacteria which contain a membrane protein called bacterio-rhodopsin, which on illumination pumps protons across the membrane. How could one exploit this protein experimentally in order to confirm that ATP synthesis in chloroplasts (and mitochondria) is driven by a proton gradient?

6. Describe how genetic information in plant cells is apportioned between the chloroplast and the nucleus. What are the implications of this apportionment? (sections 9.5.3 and 9.5.4)

9.8   Further reading

Detailed treatments of the role of chloroplasts in photosynthesis can be found in most of the standard biochemistry textbooks. Particularly recommended is chapter 19 in Alberts, B. *et al.* (1983) '*Molecular Biology of the Cell*', Garland Publishing Inc., New York.

Specialised text

Hall, D. O. and Rao, K. K. (1987). 'Photosynthesis', 4th edn, Edward Arnold (New Studies in Biology series), London.
(An excellent up-to-date account)

Reviews and articles

Arnon, D. I. (1984). *Trends Biochem. Sci.,* **9**, 258–262.
   (The discovery of photosynthetic phosphorylation)
Anderson, J. M. and Andersson, B. (1982). *Trends Biochem. Sci.,* **7**, 288–292.
   (Significance of the separation of Photosystem I and Photosystem II)
Murata, N. and Miyao, M. (1985). *Trends Biochem. Sci.,* **10**, 122–124.
   (Extrinsic membrane proteins of the photosynthetic oxygen-evolving complex)
Fluegge, U. I. and Heldt, H. W. (1984). *Trends Biochem. Sci.,* **9**, 530–533.
   (Significance of the phosphate–triose phosphate–phosphoglycerate translocator)
Schmidt, G. W. and Mishkind, M. L. (1986). *Ann. Rev. Biochem.,* **55**, 879–912.
   (Transport of proteins into chloroplasts)
Rochaix, J. D. (1985). *Int. Rev. Cytol.,* **93**, 57–93.
   (The chloroplast genome)

10 DYNAMIC INTERACTIONS BETWEEN ORGANELLES

10.1 Introduction

In chapters 3 to 9 we chose to focus on one organelle at a time, so as to concentrate on its structure, functions and biogenesis. In the process we tried to avoid engendering the view that each organelle acted in isolation: for example, we emphasised interrelationships between organelles, pointed to parallels in structure and function between them, and described the application of the same experimental techniques to their study by cell biologists. However, we appreciate that such aims may not have been achieved. In addition, the emphasis was placed firmly on the cell's proteins and the genes which encode them, at the expense of lipids and carbohydrates, which are nevertheless major cellular components. Finally, the impression may have been given that the cell, with all its organelles, is a self-enclosed entity isolated from its environment. Nothing could be further from the truth.

Let us deflect such criticism by considering further some of the dynamic interactions between organelles that, in turn, involve the plasma membrane, the interface between the cell and its environment. The overarching concept behind these considerations is that the interplay between organelles often involves an exchange of membrane. There is thus a flow of membrane around the cell's interior. This membrane traffic, mediated by transport vesicles, must be controlled and directed if the cell is to maintain its internal homeostasis. Specific aspects of the processes concerned are conveniently illuminated by the study of one well-understood receptor protein of the plasma membrane, that for low-density lipoprotein. First, however, we focus on receptor-mediated endocytosis.

10.2 Receptor-mediated endocytosis

The cell is able to internalise material from its environment by means of pinocytosis (section 6.2.2). This form of *endocytosis* is invariably non-specific: there is no selective recognition of the material to be ingested, and the delivery of the material to the lysosomes marks the end of the process. In contrast, many cells respond to signals arriving at the plasma membrane by an enhanced rate of endocytosis, with significant physiological effects. The key distinction here is that this process involves the *specific* recognition of the extracellular material by an appropriately oriented receptor in the plasma membrane.

Let us consider the details of such *receptor-mediated endocytosis* by reference to figure 10.1. The sequence of events is initiated by the interaction between the ligand and the binding site of a specific receptor (invariably a protein) in the plasma membrane (stage A). This binding process can be visualised by electron microscopy if the ligand is first labelled with the electron-dense protein, *ferritin*. Initially one sees a fairly uniform distribution of ligand, but within a few minutes or so at 37°C, the complexes of receptor-bound ligand begin to cluster at a point where a small invagination appears in the plasma membrane (stage B). The cytoplasmic face of this invagination becomes decorated with a highly organised network of clathrin (described in section 5.3.1). The result is a *coated pit* (stage C and figure 10.2). Further invagination leads to membrane fusion and budding off into the cytoplasm of a *coated vesicle* (figure 10.2), totally surrounded by a basket of clathrin and containing the ligand still bound to its receptor (stage D). Soon afterwards, the clathrin lattice disintegrates (stage E), and the individual clathrin monomers return to the plasma membrane, ready to participate in another round of endocytosis. The signal for dissociation of the

EXTRACELLULAR SPACE

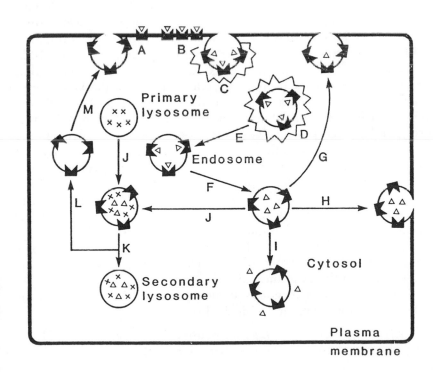

**Figure 10.1
Receptor-mediated
endocytosis**

This diagram summarises the possible fates of several different classes of external ligand in various types of cell. Details of each stage (A–M) are given in the text. [Abbreviations used: △, ligand; ◣, receptor; /\/\/\, clathrin; ×, acid hydrolase.]

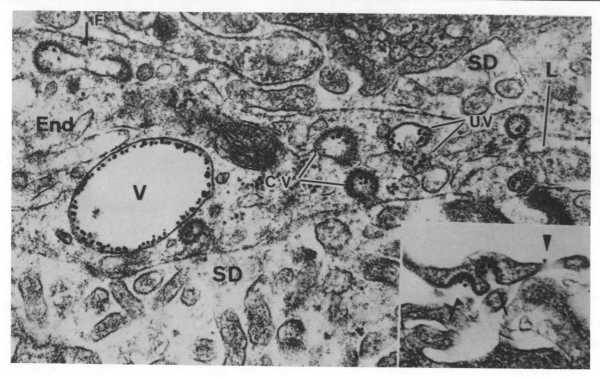

Figure 10.2
Coated pits and vesicles

Rat liver was perfused with a solution containing the protein caeruloplasmin (CP) coupled to colloidal gold particles. Within one minute this endothelial cell (End) has taken up the CP–gold (the black spots in this electron micrograph) into clathrin-coated vesicles (CV, with a 'fuzzy' outline) and uncoated vesicles (UV, with a sharp outline). Fusion between coated vesicles can be seen at the top left (flag F). A large endocytic vacuole (V) has CP–gold bound to its inner surface. SD, space of Disse. [Inset]: CP–gold has bound to the plasma membrane at two clathrin-coated pits (indicated by flags). × 48 000. [Taken from Kataoka, M. and Tavasolli, M. (1985) *J. Ultrastruct. Res.*, **90**, 194–202; reproduced by permission of Academic Press Inc.]

Endosome: Endocytic vesicles have been given numerous different names in the literature: endosomes, receptosomes, pinosomes, and phagosomes, to mention a few. The term 'endosome', which is preferred in this text, appears to be gaining the most widespread acceptance.

clathrin is not known. It may be brought about by the acidification of the matrix of the vesicle (see below).

The resulting smooth **endosome** has a number of possible fates, depending in part on the cell type. Frequently, neighbouring endosomes fuse to form large (up to 1 μm diameter) spherical, tubular or multivesicular structures lying near the cell's periphery (figure 10.3). Another early event is the progressive acidification of the endosomal contents, powered by a proton-pumping ATPase of the kind present in lysosomes (section 6.3.1) but distinct from the F_0F_1-ATPase of mitochondria (section 8.2.2). Binding of many ligands to their corresponding receptor is exquisitely sensitive to changes in pH (stage F), such that acidification by only one pH unit or so leads to release of ligand. (Compare the receptor-mediated incorporation of acid hydrolases into nascent primary lysosomes; section 6.4.3). The acidity of endosomes *in vivo* has been measured with pH-sensitive fluorescent dyes, such as chloroquine or acridine orange. Within minutes, peripheral endosomes attain an internal milieu of about pH 6.5, but the pH decreases (ultimately to about 5.0) as the endosomes migrate towards the centre of the cell and the membrane-bound H^+-ATPase continues to function.

The consequences of endosomal acidification depend on the nature of the ligand and receptor (table 10.1):

(a) *Receptor responds to low pH*: As discussed above, an acidic pH may bring about a change in the conformation of the receptor, such that

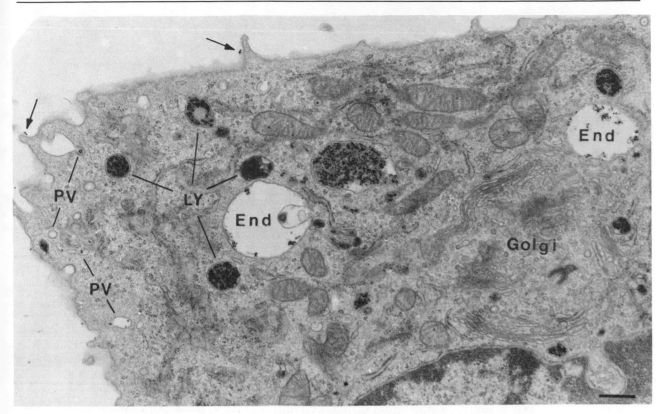

Figure 10.3
Organelles involved in endocytosis

This electron micrograph shows a mouse macrophage exposed for two hours to colloidal thorium dioxide. Electron-dense colloidal particles are adsorbed to the plasma membrane (arrows). Small (0.1–0.2 μm) pinocytic vesicles (PV) have invaginated from the plasma membrane; they have fused to form larger vacuoles called endosomes (End). Note the internal structure of the endosome at the left, and its association with tubular elements. Most of the colloidal particles are present in lysosomes (LY). The Golgi complex is not labelled. Bar = 0.5 μm; × 19 000. [Taken from Steinman, R. *et al.* (1983). *J. Cell Biol.*, **96**, 1–27; reproduced by copyright permission of the Rockefeller University Press.]

its affinity for ligand is drastically decreased. The result is the liberation of the ligand into the endosomal matrix and the retention of the receptor within the endosomal membrane. The receptor may, *via* budding of a transport vesicle, return to the plasma membrane (figure 10.1, stage G), ready to participate in another round of endocytosis. Typically, one such cycle through the cell's endocytic compartment takes about 15 minutes. In a small proportion of cases, an intact endosome may itself fuse with the apical plasma membrane and thus regurgitate the ligand.

(b) *Ligand responds to low pH*: One of the best understood examples of this effect is that of *transferrin*. This plasma protein (79.5 kDa) binds two ferric iron atoms per molecule with high affinity at neutral pH. Its receptor is a disulphide-linked dimer of 90 kDa transmembrane glycoprotein subunits, present in the plasma membrane of a wide variety of mammalian cells. At neutral pH, the tight binding between iron-loaded transferrin and its receptor stimulates endocytosis *via* coated pits. Acidification of the resulting endosome leads to a change in the conformation of the transferrin (the ligand, in this case), such that the iron is released; the apotransferrin remains bound to the receptor, however. Recycling of the endosome to the plasma membrane exposes the apotransferrin–receptor complex to an environment of neutral pH, under which conditions the complex dissociates. The apotransferrin is thereby released into the plasma. Thus, the affinity between protein

Table 10.1
Fate of receptor-bound
ligands in endosomes

Ligand	Cell type	Effect of low pH
Low-density lipoproteins, acid hydrolases	Fibroblasts	Dissociation from receptor
Asialoglycoproteins	Macrophages	Dissociation from receptor
Mannosyl glycoproteins	Hepatocytes	Dissociation from receptor
Transferrin	Many	Release of bound iron
Diphtheria toxin	Many	Membrane insertion[a]
Enveloped viruses	Many	Membrane fusion[a]

The acidic environment (pH 5.0–6.5) within the endosome causes a conformational change in the receptor (upper three examples) or in the ligand (lower three examples). [a] These processes effect release into the cytosol of toxin subunit and viral particle respectively.

and ligand (apotransferrin–iron, transferrin–receptor, apotransferrin–receptor) is precisely turned to the changes in pH that occur in this cycle of events.

Another example of ligand responding to low pH in the endosome is that of the *enveloped animal viruses*. (We discussed vesicular stomatitis virus in section 4.2.3; influenza virus is another.) When these viruses are taken into the cell by receptor-mediated endocytosis, the subsequent acidification causes a conformational change in the spike glycoprotein on the surface of their membrane envelope. The result is fusion between the membrane of the virus and of the endosome, with the liberation of the viral particle into the cytoplasm (figure 10.1, stage I). Fusion normally requires an endosomal pH of 5.1–6.0, depending on the strain of influenza virus. Certain mutant viruses, with delayed intracellular release and reduced infectivity, have amino acid substitutions in their spike glycoprotein; such mutants only induce fusion at a lower pH than does the wild-type strain.

(c) *Ligand-receptor complex does not dissociate*: Certain ligand–receptor complexes do not respond to the acidic conditions within the endosome. One fate of such complexes is their complete degradation in secondary lysosomes; for example, macrophages endocytose immune complexes by means of a surface receptor for the F_c region of **immunoglobulin G** (**IgG**) that is bound to antigen. (For an example of this process, see figure 6.12.) The resulting endosome fuses with a primary (or secondary) lysosome (figure 10.1, stage J), and the entire receptor–IgG–antigen complex undergoes degradation catalysed by acid hydrolases (stage K). This receptor is therefore not re-cycled.

In other specialised cells, the endosome may travel to the baso-lateral plasma membrane, where the ligand is released into the extracellular fluid at the opposite face of the cell (stage H). This process of *transcytosis* requires a mechanism, at present little understood, whereby the ligand-bearing vesicle avoids fusion with a lysosome before reaching the basal membrane. Whatever the mechanism, this process

IgG F_c: Immunoglobulin G has two antigen-binding sites which constitute the F_{ab} region. The 'tail' of the IgG molecule is the F_c region, which has binding sites for macrophages, complement, and other physiological effectors. When the IgG is bound to its antigen in an immune complex, the F_c region is exposed to the environment.

is exploited by several types of cell. For instance, hepatocytes take up immunoglobulin A and discharge it into the bile duct. In the intestine of newborn rats, epithelial cells take up IgG derived from the maternal milk and discharge it into the interstitial fluid at the baso-lateral surface. In the latter case, the relatively alkaline pH of the external medium may facilitate release of IgG from its receptor.

Experimental evidence confirming the significance of vacuolar acidification has come from studies with compounds which interfere with this process: weak bases (such as ammonium chloride and chloroquine) or **ionophores** (such as monensin). These compounds prevent the intracellular release of influenza virus, and reduce the rate of re-cycling of receptors to the plasma membrane. They also inhibit the delivery of acid hydrolases to nascent lysosomes, as discussed in section 6.4.3.

It seems likely that most of the re-cycled receptor is returned to the plasma membrane *via* small transport vesicles which bud from an acidified endosome (figure 10.1, stage G). Some receptors may also be selectively re-captured from secondary lysosomes by a similar process (stages L and M). Another significant aspect of this process is that it contributes membrane material itself to the cell surface. It has been estimated that professional phagocytic cells internalise 100 per cent of their plasma membrane every few hours. Clearly, there must be some compensating mechanism for replacing this internalised membrane. Although membrane biogenesis *de novo* will make a contribution (section 4.4), most of the replenishment is achieved by exocytosis of transport vesicles bearing re-cycled membrane. In most cells a small proportion of pH-sensitive receptor proteins escapes re-cycling, and they are degraded in secondary lysosomes. This *down-regulation* effect contributes to the normal turnover of such receptors. Down-regulation is enhanced when the ligand concentration increases. The result is to stabilise the physiological response to the ligand, a consequence of particular importance in the control of metabolism by peptide hormones such as insulin.

Endocytosis is, under most circumstances, balanced by exocytosis. We do not have the space here to consider further the processes of **constitutive** and **regulated exocytosis.** They were discussed briefly in relation to the role of the Golgi complex (section 5.3.1).

10.3   The receptor for low-density lipoproteins

The receptor for **low-density lipoproteins** (LDL) has been extensively studied since 1972 by **Brown** and **Goldstein.** Their interest in this particular protein concerned not only its role in the regulation of cholesterol levels in animals, but also its involvement in those clinical conditions associated with impairment of this homeostatic mechanism (for instance, heart disease). For our purposes, the life-history of the LDL-receptor vividly illustrates the interplay between numerous organelles in those cells which express this protein on their plasma

Ionophores: Compounds which insert into membranes and facilitate the passage of ions across them. For example, monensin stimulates the transmembrane exchange of H^+ ions for monovalent cations (Na^+ or K^+), and thereby reduces the acidity within endosomes.

Constitutive exocytosis: Pathway which delivers proteins and lipids to the plasma membrane, and proteins to the extracellular space. Present in all cells, and not influenced by external signals.

Regulated exocytosis: Pathway of secretion that is stimulated by specific external signals (for example, secretion of insulin by pancreatic B cells in response to a high blood-glucose level).

Low-density lipoproteins: One class of lipid–protein aggregate present in the blood plasma of animals. These spherical particles (diameter 22 nm, mass 3000 kDa) contain a core of cholesteryl ester surrounded by a shell of cholesterol and phospholipid, together with one copy per particle of apoprotein B-100.

Figure 10.4

Life-history of the mammalian LDL-receptor

Expression of LDL-receptor at the plasma membrane follows the route of a typical secretory protein, from the endoplasmic reticulum *via* the Golgi complex. The membrane-bound receptor re-cycles *via* coated pits, coated vesicles, endosomes, and re-cycling vesicles. Small vertical arrows denote the direction of regulatory effects. LDL, low-density lipoprotein; HMG CoA reductase, 3-hydroxy 3-methyl glutaryl coenzyme A reductase; ACAT, acyl CoA:cholesterol acyltransferase. [Taken from Brown, M. S. and Goldstein J.L. (1985). *Current Topics in Cellular Regulation*, vol. 26 (Levine, R. L. and Ginsburg, A., eds), pp. 3–16, Academic Press Inc., Orlando; reproduced by permission of the publisher.]

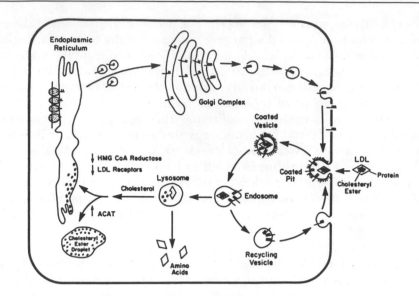

Figure 10.5

The LDL-receptor: a 'mosaic' protein

(a) The human gene for the LDL-receptor is split by 17 introns of variable lengths. Mutations in the gene are of four types: insertion (▼); nonsense (●); missense (■); and deletion (↔). kb, kilobases; nt, nucleotides. (b) The organisation of exons in the gene corresponds to the arrangement of the six functional domains in the receptor protein precursor. The ligand-binding domain contains severn internal repeat sequences (I–VII), and shows homology to the C9 protein of the complement cascade. The largest extracellular domain contains three cysteine-rich repeats (A, B, C), and shows homology to epidermal growth factor (EGF). The positions of introns are indicated (▼). [(a) taken from Brown, M. S. and Goldstein, J. L. (1986). *Science*, **232**, 34–47; copyright 1986 by The Nobel Foundation. (b) taken from Suedhoff, T. C. *et al.* (1985). *Science*, **228**, 815–822; copyright 1985 by the AAAS.]

membrane. An overview of the interactions involved is presented in figure 10.4.

The gene for the LDL-receptor, present on chromosome 19 in the nucleus of a human cell, extends over 45 kb of DNA, and each exon encodes a functional domain in the protein (figure 10.5). For example, exon 1 specifies the signal sequence which directs the nascent polypeptide to the lumen of the rough ER, as described previously for the synthesis of secretory proteins (section 4.2.2). The completed, membrane-bound

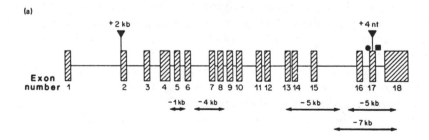

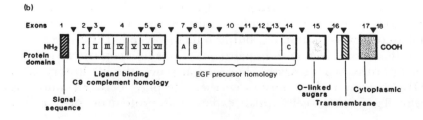

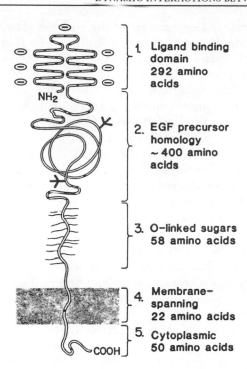

1. Ligand binding domain
   292 amino acids

2. EGF precursor homology
   ~ 400 amino acids

3. O-linked sugars
   58 amino acids

4. Membrane-spanning
   22 amino acids

5. Cytoplasmic
   50 amino acids

Figure 10.6
Structure of the human LDL-receptor

Most of the mature protein (about 750 residues) extends into the extracellular fluid, leaving a short (50 residues) cytoplasmic 'tail'. The positions of the *N*-linked oligosaccharide chains (⊀) and of the *O*-linked sugars, are indicated. The signal sequence of the precursor protein has been removed. EGF, epidermal growth factor. [Taken from Brown, M. S. and Goldstein, J. L. (1986). *Science*, **232**, 34–47; copyright 1986 by The Nobel Foundation.]

receptor polypeptide undergoes *N*-linked glycosylation at two asparagine residues, and the first sugar (*N*-acetylgalactosamine) of each of the 18 *O*-linked oligosaccharide chains is added to a serine or threonine residue. Transport vesicles ferry the receptor to the Golgi complex. Here the *N*-linked oligosaccharides are converted to complex-type chains (as depicted in figure 5.8), while galactose and sialic acid are added to create the *O*-linked chains. About 30 minutes after the start of its synthesis, the mature LDL-receptor (apparent molecular mass 160 kDa) is complete (figure 10.6). It is ferried in transport vesicles to the plasma membrane, where it appears after a further 15 minutes.

At the cell surface, the receptors (either free or with bound LDL from the plasma) cluster into coated pits and are internalised as described for receptor-mediated endocytosis. At the acidic pH within the endosome, LDL dissociates from its receptor and is delivered to a lysosome. The receptor re-cycles to the plasma membrane, ready for another round of endocytosis. Each such journey takes about 10 minutes, and occurs whether the receptor has bound LDL or not; each receptor molecule undergoes about 100 such journeys before its degradation.

In the lysosome the apoprotein component of the LDL is degraded by proteases, and acid lipase liberates free cholesterol from the cholesteryl esters. One effect of the LDL-derived cholesterol is to inhibit 3-hydroxy 3-methylglutaryl coenzyme A reductase (*HMG CoA reductase*), the rate-limiting enzyme in the pathway of cholesterol biosynthesis in the endoplasmic reticulum. Here is a good example of negative feedback. The inhibitory effect is achieved in two ways: by suppressing the transcription of the gene for HMG CoA reductase, as well as by enhancing the rate of degradation of the enzyme. The free cholesterol has two additional complementary influences. It stimulates acyl CoA: cholesterol acyltransferase (ACAT) so that excess cholesterol can be stored as cytoplasmic droplets of cholesteryl ester. It also reduces the number of LDL-receptor molecules in the plasma membrane by decreasing the concentration of receptor mRNA available for translation. The sum total of all these effects is to regulate the level of free cholesterol in the cell within narrow limits.

Hypercholesterolaemia: Elevated levels of plasma cholesterol.

The importance of this receptor-mediated pathway for cholesterol homeostasis is demonstrated by several clinical conditions associated with **hypercholesterolaemia.** Four classes of inherited defect in the gene for the LDL-receptor have been described (figure 10.5a and table 10.2). The best understood examples (the Class 4 mutations) have been studied at the molecular level by sequencing cloned DNA derived from genomic libraries. In each case, the genetic lesion affects the *C*-terminal tail. How does this region of the LDL-receptor interact with coated pits in order to bring about internalisation? We do not know, but this interaction is currently attracting considerable attention from cell biologists.

In conclusion, the life-history of the LDL-receptor dramatically illustrates the interplay between numerous organelles: the nucleus, endoplasmic reticulum, Golgi complex, endosomes and lysosomes, as well as the plasma membrane and several types of transport vesicle. We have not mentioned here the mitochondrion, peroxisome or chloroplast. However, a good example of metabolic cooperation between these three organelles is afforded by photorespiration in plant cells, as described in section 7.2.3.

10.4   Future perspectives

Which questions should cell biologists be asking as they plan their future research? Where is progress likely to be made in studies of the structure, function and biogenesis of organelles, and the interactions between them? What light will recently developed techniques shed on these areas of interest? Advances will probably come in two major spheres of investigation.

Table 10.2
Mutations in the gene for the LDL-receptor

Class of mutation	Synthesis	Transport from ER to Golgi	Binding of LDL	Clustering in coated pits
1	X			
2	⟶ X			
3	⟶ X			
4	⟶ X			

In the life-history of the LDL-receptor, four classes of mutation (each in a different region of the gene) disrupt its function at the site indicated by the cross.

First, consider the genetic aspects. Cloning and sequencing of genes are already commonplace, as is the identification of gene loci on chromosomes by *in situ* hybridisation (see, for example, figure 2.14). In between these fine and coarse levels of DNA analysis, there was until recently no experimental technique for studying fragments of genetic material in the size range 50–1000 kb. Pulsed-field gel electrophoresis now provides this information (for an example of this technique, see figure 2.9). This method will simplify the characterisation of extended segments of genomic DNA containing structural genes *and* their associated regulatory elements. Indeed, cell biologists will focus increasingly on the means by which the cell controls the expression of the protein components of its organelles. For instance, why is urate oxidase present in the peroxisomes of rat liver but not of other tissues, and not in other species? How does light activate the synthesis of specific chloroplast proteins? Back at the fine level, the automation of DNA sequencing will provide ever more rapidly the evidence for homology between functionally related genes (or exonic segments of protein-coding genes). Such information will point to further phylogenetic links between those organelles (or their components) with a common evolutionary ancestor.

Secondly, consider the cellular aspects. We already have firm evidence that many organellar proteins contain targeting signals in their structure that interact with specific receptors and processing enzymes. We need to reconstitute these recognition processes *in vitro*, and to determine how each receptor/processing system comes to be organised in its corresponding organelle. Then there is the major topic of transport vesicles. These membrane-bound structures ferry material around the cell: from endoplasmic reticulum to the Golgi complex; from one Golgi saccule to another; from that organelle to the plasma membrane, which also receives re-cycling vesicles from endosomes. How do all these vesicles form, and how is their subsequent fate determined? The processes involved need to be reconstituted *in vitro* so that their components can be dissected. Further information on this point will be derived from studies of appropriate genetic mutations in simple eukaryotic organisms which can be experimentally manipulated, such as yeast. Finally, the interactions between organelles in living cells deserve greater attention. Video-based cinematography with image enhancement will provide such images; in addition, conventional microscopes fitted with charge-coupled devices will generate data from which computer analysis can project three-dimensional images of the cell's interior.

These are some of the exciting experimental approaches of the not-too-distant future.

10.5   Summary

Receptor-mediated endocytosis internalises specific ligands from the cell's external environment. Clathrin-coated pits and vesicles give rise to endosomes. This pre-lysosomal compartment undergoes acidification by a membrane-bound H^+-ATPase. In some cases the relatively low pH induces a conformational change in the receptor, which releases its ligand and returns to the plasma membrane. In other cases the ligand may undergo an acid-induced change in structure, or the receptor–ligand complex may remain unmodified. The LDL-receptor is involved in cholesterol homeostasis. It delivers lipoprotein-associated cholesterol to the cytoplasm in a tightly regulated process which matches the cell's demand for this lipid. Mutations in the gene for the LDL-receptor block its synthesis, its transport to the plasma membrane, its binding of LDL, or its internalisation. The pathways of cholesterol homeostasis involve interactions between components of numerous organelles. Future research in cell biology is likely to focus on genetic control of the structure, function and biogenesis of organelles, and on the reconstitution of vesicular traffic within the cell.

10.6   Further reading

Dautry-Varsat, A. and Lodish, H. F. (1984). *Sci. Amer.,* **250** (5), 48–54.
   (Receptor-mediated endocytosis)
Helenius, A., Mellman, I., Wall, D. and Hubbard, A. (1983). *Trends Biochem. Sci.,* **8**, 245–250.
   (Endosomes)
Mellman, I., Fuchs, R. and Helenius, A. (1986). *Ann. Rev. Biochem.,* **55**, 663–700.
   (Endocytosis and exocytosis)
Brown, M. S. and Goldstein, J. L. (1986). *Science,* **232**, 34–47.
   (The LDL-receptor story)

APPENDIX A: ABBREVIATIONS

Amino acids

Ala, alanine (A)
Arg, arginine (R)
Asn, asparagine (N)
Asp, aspartic acid (D)
Cys, cysteine (C)
Gln, glutamine (Q)
Glu, glutamic acid (E)
Gly, glycine (G)
His, histidine (H)
Ilc, isoleucine (I)
Leu, leucine (L)
Lys, lysine (K)
Met, methionine (M)
Phe, phenylalanine (F)
Pro, proline (P)
Ser, serine (S)
Thr, threonine (T)
Trp, tryptophan (W)
Tyr, tyrosine (Y)
Val, valine (V)

Nucleotide bases

A, adenine; C, cytosine
G, guanine; T, thymine
U, uracil

Nucleotides

AMP, adenosine 5'-monophosphate
ADP, adenosine 5'-diphosphate
ATP, adenosine 5'-triphosphate
(or any of the other bases in place of A)

Nucleic acids

DNA, deoxyribonucleic acid
RNA, ribonucleic acid
mRNA, messenger RNA
rRNA, ribosomal RNA
tRNA, transfer RNA
cDNA, complementary (copy) DNA
bp, base pairs
kb, kilobases (1000 nucleotides)

General

EM, electron microscopy
ER, endoplasmic reticulum
kDa, kiloDalton (the commonly used unit of molecular mass)
P_i, inorganic phosphate
PAGE, electrophoresis in polyacrylamide gel
SDS, sodium dodecyl sulphate

APPENDIX B: ANSWERS TO STUDY QUESTIONS

Chapter 1

1. Refer to table 1.2.
2. About 6 per cent.
3. Acetyl CoA is converted by the mitochondrion into a metabolite for which there *is* a transport protein, in this case citrate. Cytosolic citrate is cleaved enzymically to regenerate acetyl CoA.
4. The peroxisome contains those enzymes that *generate* H_2O_2 (for details, see figure 7.1).
5. The lysosomal membrane probably contains a proton pump to acidify the organelle's interior; transport proteins to allow the exit of the polar digestion products; receptor proteins to recognise legitimate targets for fusion; there are also membrane-bound enzymes.
6. Incubate a tissue section with a specific substrate which one of the acid hydrolases can convert to an electron-dense product (see, for example, figure 6.4).

Chapter 2

1. Freeze-fracture EM and circular dichroism studies confirmed that membranes contained integral proteins with globular (non-extended) conformations.
2. Ion-exchange chromatography would separate Protein C from Proteins A + B; gel filtration would resolve Protein A from Protein B.
3. Possible explanations include: the peroxisomal membrane is permeable to the substrate used; the peroxisomal membrane has been damaged and made 'leaky'; the active sites of the enzymes concerned are exposed to the external environment.
4. Probably, sheets of plasma membrane sediment with nuclei, and vesicles (generated by shearing the plasma membrane) sediment with microsomes.
5. Refer to figure 2.13. Use an oligonucleotide probe (derived from

the amino acid sequence of GH) to recognise GH-cDNA. Use an expression vector system to generate GH.

6. The polyclonal antibody recognises a common epitope (on the α subunit); the monoclonal antibody recognises an unique epitope (on the γ subunit).

7. Couple convanavalin to agarose beads; add derivatised beads to (glyco)protein solution; centrifuge, recover beads with bound glycoproteins; elute glycoproteins with high concentration of α-D-mannose.

Chapter 3

1. About 1600 vesicles.

2. Refer to section 3.4.2.

3. By X-ray diffraction of crystallised octamers; by experimentally cross-linking adjacent monomers, and identifying the components of the resulting dimers etc.

4. See if the precursor lamin A, but not the mature protein, is imported by nuclei *in vitro*; fuse the 2 kDa extension polypeptide to a cytosolic marker protein and look for nuclear import; coat colloidal gold particles with the 2 kDa polypeptide and look for nuclear import (see figure 3.11).

5. Coat colloidal gold particles with oligo (dT), which should complex with the poly(A) + tail of mRNA; micro-inject the particles into the nucleus of a cultured cell; look (by EM) for particles passing through nuclear pores.

6. Nucleolar DNA comprises mainly genes for rRNA (see figure 3.13). Radioactively labelled RNA would hybridise to these genes *in situ* if incubated with suitably prepared chromosomes.

7. Lipids and proteins of the outer nuclear membrane are probably incorporated from the ER; lipid-transport proteins may insert lipids into the nucleoplasmic face of the inner membrane, which probably has specific receptors that incorporate newly synthesised proteins; the nuclear lamina may have temporarily disorganised regions where new lamin molecules can be integrated.

Chapter 4

1. Refer to table 4.2.

2. Refer to section 4.4. In summary, a pre-existing, asymmetric membrane template is required.

3. Hydrolysis of PIP_2 generates diacylglycerol (DAG) and inositol trisphosphate (IP_3). The effects of the DAG-stimulated protein kinase would augment or complement those of the Ca^{2+}-dependent protein kinases activated by Ca^{2+} released by IP_3.

4. *In vitro* translation of mRNA for ovalbumin generated a polypeptide with the same size as the mature protein.

5. Label pre-existing ('old') membranes metabolically with [^{32}P]-phosphate; after phenobarbital induction, isolate smooth ER; if

the new membrane was acquired from other intracellular sites, the enlarged smooth ER would have the same specific radioactivity as that before induction; newly synthesised membrane would lead to a decrease in specific radioactivity.

6. Assay lipid vesicles, reconstituted with signal peptidase complex and docking protein, for translocation competence (that is, uptake of secretory protein into a protease-resistant form).

7. Incubate VSV-infected cells with radioactive palmitate; fractionate cell homogenates by centrifugation; assay fractions for marker proteins and radioactive G protein.

Chapter 5

1. Refer to table 5.1 and section 5.2.1.
2. Refer to figure 5.6.
3. Presumably newly synthesised galactosyl transferase has a targeting signal recognised by a receptor protein which itself is a resident of the *trans* cisternae.
4. Analyse the composition of the oligosaccharide chains of the glycoprotein acid hydrolases. If they contain complex-type chains, they must have passed through the *trans* sub-compartment of the Golgi complex. (In fact, this has been shown to be the case.)
5. Endocytic vesicles must have fused with Golgi cisternae, thereby allowing ASOR to be acted on by sialyl transferase in the *trans* cisternae. Exocytosis releases sialylated ASOR.
6. Isolated Golgi vesicles should contain protease-resistant glycosyl transferase activity; these vesicles should not catalyse glycosylation of exogenous protein substrates.

Chapter 6

1. Refer to sections 6.1 and 2.2.1.
2. Refer to figure 6.4.
3. Activity is only manifest when the lysosomal membrane is disrupted.
4. Acid hydrolases synthesised by the rough ER are incorporated into specific vesicles (nascent lysosomes) by the Golgi complex.
5. Briefly, pulse-chase labelling has identified precursors of nascent acid hydrolases; cDNA cloning has revealed their amino acid sequences; and immuno-EM has indicated the intracellular site of biogenesis of lysosomes.
6. Briefly, a mutation in the gene for an acid hydrolase results in inactive (or no) enzyme; the undegraded substrate(s) of that enzyme accumulate(s) in secondary lysosomes.
7. About 2.5×10^6 molecules per lysosome (assuming an even distribution of enzyme); about 100 mg of protein per ml.

Chapter 7

1. By histochemical staining for catalase (peroxisomes) or acid phosphatase (lysosomes).

2. mRNA from control rat liver should hybridise with and thereby remove cDNA corresponding to non-peroxisomal proteins in Clofibrate-treated liver.

3. Carnitine octanoyl- and acetyl-transferase (for exporting products of peroxisomal β-oxidation); H^+-ATPase (acidifying peroxisomal matrix); permeases for phosphoglycolate and glycine, and for α-oxoglutarate and glutamate (for photorespiration in plants).

4. By immuno-EM with a monoclonal antibody to the 22 kDa peroxisomal membrane protein.

5. Loss of photorespiration metabolism (accumulation of glycolate); impaired degradation of chlorophyll (accumulation of phytanic acid); accumulation of very-long-chain fatty acids.

Chapter 8

1. All those associated with mitochondria (in particular, Krebs cycle, β-oxidation, ketone body metabolism, haem synthesis, electron transport chain).

2. Mitochondrial translation would be blocked by chloramphenicol; cytoplasmic translation (of nucleus-encoded mRNA) would be blocked by cycloheximide.

3. Refer to the relevant sections.

4. Impaired oxidation of NADH by electron-transport chain; increase in $NADH/NAD^+$ ratio in mitochondria; inhibition by NADH of pyruvate dehydrogenase and Krebs cycle enzymes; inadequate levels of ATP.

5. Most likely, such sequences are not exposed in cytosolic proteins, and so are not accessible to the mitochondrial receptor proteins.

6. Use affinity chromatography, with detergent-solubilised proteins of the outer mitochondrial membranes and either insolubilised pre-sequence for apocytochrome c, or insolubilised antibody to the 45 kDa protein.

7. Noradrenaline (NA) releases a second messenger which binds to the 32 kDa protein and thereby uncouples the mitochondria; the free energy of electron transport is lost as heat. (In fact, the second messenger is known to be free fatty acid, released by NA-stimulated hydrolysis of triglyceride.)

Chapter 9

1. 20.

2. A polypeptide of 30 kDa contains about 250 amino acids, which would be specified in the mRNA by 250 codons, or 750 bp (0.75 kb). 120 such polypeptides would thus require 90 kb of DNA, much less than the size of the average chloroplast genome.

3. It would reduce the rate of CO_2 fixation (see section 7.2.3).

4. Photosystems I and II; cytochrome b/f complex; CF_0CF_1 ATP synthase; light-harvesting chlorophyll–polypeptide complexes.

5. Incorporate bacteriorhodopsin into artificial lipid vesicles, together with ATP synthase from chloroplasts (or mitochondria); assuming correct orientation of the proteins, illuminated vesicles should generate a proton gradient which results in detectable synthesis of ATP.

6. Refer to table 9.2 and figure 9.14. Nucleus-encoded chloroplast proteins, synthesised on cytoplasmic ribosomes, must contain targeting sequences that ensure their integration into the correct compartment of the organelle and with the correct orientation.

INDEX

Page entries in normal typescript are to references in the text. Those in *italic* are to figures and those in **bold** are to tables.